THIS BOOK BELONGS TO:

_____

FUNDAMENTALS OF FOOD PROCESSING AND TECHNOLOGY

By

Wilbur A. Gould, Ph.D.

Emeritus Professor,
Food Processing and Technology,
The Ohio State University
and
Executive Director,
Mid-America Food Processors Association
and
Consultant to the Food Industries

FUNDAMENTALS OF FOOD PROCESSING AND TECHNOLOGY

A technical reference book and textbook for students of food technology, food plant managers, product research and development specialists, food brokers, technical salesmen, food equipment manufacturers, and food industry suppliers.

Library of Congress Catalog-in-Publication Data

Gould, Wilbur A., 1920-
   Fundamentals Of Food Processing And Technology,
   by Wilbur A. Gould.
   p.   cm.
   Includes Index

   1. Food Industry and Trade.   I. Title.
TP370.G65   1997
664 - - dc21   97-20766
                                                              CIP

While the recommendations in this publication are based on scientific studies and industry experience, references to basic principles, operating procedures and methods, types of instruments and equipment, and food formulas, are not to be construed as a guarantee that they are sufficient to prevent damage, spoilage, loss, accidents or injuries, resulting from use of this information. Furthermore, the study and use of this publication by any person or company is not to be considered as assurance that that person or company is proficient in the operations and procedures discussed in this publication. The use of the statements, recommendations, or suggestions contained, herein, is not to be considered as creating any responsibility for damage, spoilage, loss, accident or injury, resulting from such use.

A PUBLICATION OF

CTI PUBLICATIONS, INC.
2 Oakway Road, Timonium, Maryland 21093-4247USA

PREFACE

This book was written to summarize some of the fundamentals to be considered in the food processing and technology area. It is an outgrowth of my Workshop on this subject, which has become quite popular with personnel working in food firms.

It all started with my introduction to the subject when I was a student in the late '30s and early '40s. I taught night school while in the U.S. Navy in Pearl Harbor, and this was one subject many men developed an interest for. After the war and my joining the faculty at The Ohio State University in 1946, it was the basis for my introducing students to the food processing and technology field. It became a natural course of instruction to recruit students to this subject area, and it proved to be a great course of study to show students the past, the present and the concerns we have today for some insight into the future.

I make no apologies for my shortening this subject, as my hope is that students would seize the opportunity to go further with many aspects of this subject. I have many friends and former students who tell me this was their starting point, moving forward from here. As a matter of fact, one former students claims this was the turning point in his life, and just this past summer he willed to The Ohio State University $1.5M to establish a food Chair in my name for what little I did for him. One never knows what turns people around. I know in my life the introduction of this subject changed my interest from plant breeding to food processing and technology. I am deeply grateful for the change, and my hope is that those who read and study this book will be so motivated to move further in this direction.

The industry needs people who are motivated to help process, preserve, create and develop newer technologies. Man must eat, and we in this profession can and will make significant contributions to provide food for all concerned, as we work toward better and significantly improved practices. The challenge is to do like many of the early contributors whom I have cited. The challenge is to produce food that we need. The challenge is to improve quality. The challenge is to improve efficiency, and the challenge is to develop new businesses that add value to the preserved product. We know it can be done and that there will still be many more pioneers, entrepreneurs, food technologists and scientists who want to make their contributions. We all have an obligation to produce, pack and preserve the best food for all to enjoy. We must do it efficiently and profitably, and we must always put safety first.

— **Wilbur A. Gould**

ACKNOWLEDGEMENT

Writers usually have many friends, peers and associates to acknowledge. I have acknowledged many in my other books, but, with this book, I wish to acknowledge my family since they have been my inspiration, my reason for delving into this activity, and my true love—because they all believe in what I have been doing.

Jessie, my first real love, true partner and dearest friend, has read all my writings and has toned them down, corrected my misuse of the English language, and generally given me insights and help to pull together my efforts. She leaves me alone when I am at work, but she takes the time to check out the details when asked. She has been much more than a wife, a mother and an associate. To me, she is my best friend, my best critic, and my best mentor. She deserves much more credit than anyone could give her for being just who she is.

Ronald, our oldest, has been a major source of inspiration and a real confidant. He is very frank and tells it like it is. He never divulges secrets from his job with General Mills, but he challenges me to cover certain subjects because they are in use every day. He is one of the first to give courage, pass judgment and support, whenever and wherever possible. I owe a deep debt of gratitude to him for his listening ability, his constructive criticism and his willingness to go the extra mile to help. He is a real food technol-ogist, food engineer and a true practicing food scientist.

Rebecca, our second child, learned early in life that she did not want science and chose the business field. Our discussions of the business end of any industry have been constructive and frank. She operates a successful business in Australia, and it is a great break for us to go and see how she and her husband enjoy life and manage their affairs. She was my first secretary at OSU and she edited my first book; she has done much of her own writing since.

Jacquelyn, our third child, is a very successful recruiter in the Chicago area with her own firm, The Career Group, responsible for recruiting professionals for the food industry. Her experiences with the Kroger Co., General Mills and Sara Lee Corp. honed her skills while working with people, and developed her personality to meet and greet and judge people. She and her husband participate in many food shows, seminars and technology activities. Her efforts have proven successful in recruiting people for many food firms, and they far overshadow my former recruiting and placement activities.

To say the least, my family and my two grandsons, Ronald A. and Brian R., are all making their contributions and, hopefully,   the food industry will be better off for the many efforts of The Goulds.

My sincere Thanks to Randy and Nancy Gerstmyer, and Art Judge II, all of CTI Publications, for their help and encouragement in publishing this and other works of mine. They are true communicators for this industry.

— Wilbur A. Gould

CONTENTS

FUNDAMENTALS OF FOOD PROCESSING AND TECHNOLOGY

Chapter 1

INTRODUCTION

To survive, man must eat.
To maintain good health, man must eat right.

The food industry is the most important industry in the world. without food, man could not survive. The food industry is big business with well over $300,000,000,000.00 of annual sales each year. The food industry employs one out of every seven working persons. The food industry is a low margin industry, but the work is steady and most necessary. The food industry requires knowledge of microorganisms and general sanitation practices for most employees. It is an industry that is changing and improving in the training and education of all personnel. One economist has recently predicted that college educated people will be more common within the industry before the next century due to computers, technology, and greatly expanded sciences and engineering know how.

Each one of us eat approximately 1400 pounds of food per year, yet, our food in the USA is the cheapest in the world—as less than 14 cents of the earned dollar is actually spent on food. About half of the food we eat is from animal sources and the other half is from plant sources. However, the cost of our food is about 40 cents out of the earned dollar for plant products and some 60 cents for the animal products. Times are changing as more and more plant foods are replacing animal foods.

Food Composition

How much we eat is only one criteria to describe the food business. More importantly is what we eat in terms of nutritional values. Food serves many functions, that is, it provides energy (calories) to sustain life, food promotes growth, food is needed to repair damaged and worn out tissues, food is essential for reproduction and food provides unique esthetic and psychological values. Food is most important to our well being.

Food is composed of water, carbohydrates, proteins, fats, minerals, vitamins, organic acids, enzymes, hormones, pigments, flavor compounds and many other constituents. The combination of these compounds is what makes foods different. In some cases the combination of nutritional compounds is effected by the modern food processor to meet the demands of customers.

Water is the primary constiuent of many foods and it is essential along with carbon dioxide, the green plant and sunshine for food production for man and animals to live on. My former professor, Dr. Fred Deatherage, used to start his lecture on water by stating "there is no life without water". Water makes up a large portion of most foods and water is vital for living cells and/ or all biological functions.

The rate at which biological and chemical activities takes place is controlled in large part by temperature. If we change the temperature of a product by 10°C. (18°F.), we double the rate of reaction. For example, if the normal shelf life temperature is 50°C. and we take the temperature up to 60°C. we can assume a doubling of the rate of reaction. Likewise, if we lower the temperature 10°C. we will slow the rate of reaction in half. This is known as the Q sub 10 rule and very vital to understanding changes in rates of chemical and biological reactions.

Understanding of Food Labels and Labeling

In May of 1994, labeling of food changed requiring the food processor to add nutritive information to inform the consumer about the nutrient quality of the food in the package. This information is most relevant. For examples, calories inform us with information relative the energy in the food. It is stated as an amount per serving The percent from fat, protein or carbohydrates are declared and the percentage of the Daily Values based on a 2000 calorie diet is given. Some people will require more or less than 2000 calories. Some atheletes may consume 10,000 calories or more per day. One can calculate the number of calories consumed by knowing the amount of food eaten in grams and then reading the label to determine the percent from fat, carbohydrate or protein and multiplying the grams of fat by 9, and grams of carbohydrates by 4 and the grams of proteins by 4 and totaling these to determine the total calories. For example, the label shows us that an average serving is 200 grams and the food has 100 grams of carbohydrates, 50 grams of fat, and 50 grams of protein, therefore the total calories equals 1050. All consumers should become knowledgeable of and about food nutrients and the labels on foods and the nutritional information they provide. For example, no more than 30% of an individuals calories should come from the fat in the daily diet.

Another important aspect of our food is what we eat. No single food can supply all the nutrients the body needs. Generally, we should eat from each of the food groups as shown in the Food Pyramid, that is, we should eat 6 to 11 servings per day from the bread, cereal, rice and pasta group; 2 to 4 servings per day from the fruit group; 3 to 5 servings per day from the vegetable group; 2 to 3 servings per day from the meat, poultry, fish, dry bean, egg or nuts group; 2 to 3 servings per day from the milk, yogurt and cheese group, and very sparingly from the fats, oils, and sweet group. These groups have been developed by the U.S. Department of Agriculture and are illustrated in the Food Guide Pyramid (see Figure 1.1).

The labeling of foods tells us the percent of the Daily Values (DV). The DV's help us to compare one food to another and whether or not a food makes a contribution to our diet. Figure 1.2 is an example of the labeling for a given type of cookies. Reading the labeling and using the information can be most helpful to all consumers.

Labeling of food is a valuable tool in terms of nutrients, however, food processors and their food products will usually be different in terms of various aspects of quality. Quality is a major criteria of all food products. Some processed products will have better color, texture, shape, appearance, flavor and freedom from defects than other products. The customer should read the label in terms of brands and grade differences that may be displayed. Foods are no different than people, cars, clothing etc. as differences exist even though the nutrient qualities may be the same. Yes, even the canned Vs the frozen Vs the dried etc. for any given commodity may be different and the smart consumer learns to take advantage of these differences.

Improvements in processing technology continue to be made within the industry for the ultimate benefit of the customer.

FIGURE 1.1 — Food Guide Pyramid
(A Guide to Daily Food Choices)

KEY
◼ Fat (naturally   ▼ Sugars (added)
   occurring and added)
These symbols show that fat and added sugars come mostly from fats, oils and
sweets, but can be part of or added to foods from the other food groups as well.

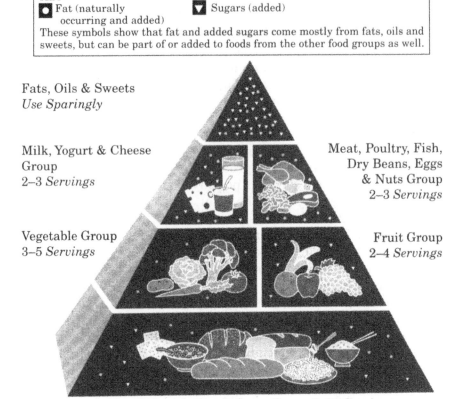

Fats, Oils & Sweets
Use Sparingly

Milk, Yogurt & Cheese
Group
2–3 Servings

Meat, Poultry, Fish,
Dry Beans, Eggs
& Nuts Group
2–3 Servings

Vegetable Group
3–5 Servings

Fruit Group
2–4 Servings

Bread, Cereal, Rice & Pasta Group 6–11 *Servings*

HOW TO USE THE DAILY FOOD GUIDE

Breads, Cereals, Rice, Pasta
1 slice bread
$^1/_2$ cup cooked rice or pasta
$^1/_2$ cup cooked cereal
1 oz. ready-to-eat cereal

Vegetables
$^1/_2$ cup chopped raw or cooked
   vegetables
1 cup leafy raw vegetables

Fruits
1 piece fruit or melon wedge
$^3/_4$ cup juice
$^1/_2$ cup canned fruit
$^1/_4$ cup dried fruit

Milk, Yogurt, Cheese
1 cup milk or yogurt
$1^1/_2$ to 2 oz. cheese

**Meat, Poultry, Fish,
Dry Beans, Eggs, Nuts**
$2^1/_2$ to 3 oz. cooked lean meat,
   poultry or fish
Count $^1/_2$ cup cooked beans, or
   1 egg, or 2 tbs. peanut butter
   as 1 oz. lean meat (about $^1/_3$
   serving)

Fats, Oils, Sweets
*Limit calories from
these*, especially if
you need to lose
weight

*The amount you eat may be more than
one serving. For example, a dinner
portion of spaghetti would count as two
or three servings of pasta.*

FIGURE 1.2 — Package Labeling for
A Given Type of Cookie

Serving Size
reflects the amount
typically eaten by
many people.

The list of nutrients
covers those most
important to the
health of today's
consumers.

COOKIES

Nutrition Facts

Serving Size 3 cookies (34g/1.2 oz)
Servings Per Container About 5

Amount Per Serving

Calories 180   Calories from Fat 90

% Daily Value*

Total Fat 10g	**15%**
Saturated Fat 3.5g	**18%**
Polyunsaturated Fat 1g	
Monounsaturated Fat 5g	
Cholesterol 10mg	**3%**
Sodium 80mg	**3%**
Total Carbohydrate 21g	**7%**
Dietary Fiber 1g	**4%**
Sugars 11g	
Protein 2g	

Vitamin A 0%	•	Vitamin C 0%
Calcium 0%	•	Iron 4%
Thiamin 6%	•	Riboflavin 4%
Niacin 4%		

* Percent Daily Values are based on a 2,000
calorie diet. Your daily values may be higher
or lower depending on your calorie needs:

		Calories	2,000	2,500
Total Fat	Less than		65g	80g
Sat Fat	Less than		20g	25g
Cholesterol	Less than		300mg	300mg
Sodium	Less than		2,400mg	2,400mg
Total Carbohydrate			300g	375g
Dietary Fiber			25g	30g

Ingredients: Unbleached enriched wheat flour
[flour, niacin, reduced iron, thiamin mononitrate
(vitamin B₁)], sweet chocolate (sugar, chocolate
liquor, cocoa butter, soy lecithin added as an
emulsifier, vanilla extract), sugar, partially hydro-
genated vegetable shortening (soybean, cotton-
seed and/or canola oils), nonfat milk, whole eggs,
cornstarch, egg whites, salt, vanilla extract,
baking soda, and soy lecithin.

Calories from Fat
are now shown on
the label to help
consumers meet
dietary guidelines
that recommend
people get no more
than 30 percent of
the calories in their
overall diet from fat.

% Daily Value (DV)
shows how a food
in the specified
serving size fits into
the overall daily
diet. By using the
%DV you can easily
determine whether
a food contributes a
lot or a little of a
particular nutrient.
And you can
compare different
foods with no need
to do any calcula-
tions.

FIGURE 1.3 — Anatomy Of A Food Label

Today's food is the highest in variety and quality it has ever been due to existing technology being utilized to its fullest. The consumer is the beneficiary if they learn to read the label and follow the simple directions as illustrated in the food pyramid.

Why Process Foods?

We have become very efficient in agriculture as today one farmer produces enough food to feed over 90 people. Much of what we produce and process is used to feed many people in other parts of the world. Food production, processing, and marketing is a big part of our economy.

At the present time we process well over 90% of the food we eat. A large portion is canned, but significant portions are frozen, fermented, dried, fried, and/or extruded. Processing of our foods makes them stable for around the year consumption. Food processing, also, makes our food convenient, safe, economical, nutritious, and inexpensive. Processing of our foods prevents spoilage and eliminates waste. Processing of our food provides vareity, year around supplies, maintenance of high and uniform quality, and foods that are good for us. Processed foods are highly acceptable by all levels of society.

Further, the processed foods industries makes significant contribution to the economy through jobs, new businesses and, of course, adding much value to the original raw products. Processed foods are an essential part of this great Nation and should be to all the world. In many parts of the world, food processing is only a developing industry and it is usually a "feast or famine" situation. That is, much food is available at the time of harvest or kill and due to lack of presevation it is not preserved and, thus, only consumed during the production season(s).

The food industries constantly seeks young men and women that know the profession of food processing, science, and technology. Opportunities are commensurate with ones experiences, desires, education, and ability to grow and contribute to his or her chosen firm. The challenge is greater today than ever before for bright young minds to excel and contribute and move to the forefront as the leaders of tomorrow. The industry needs these bright young minds and their many contributions.

Nearly every Land Grant College or University and many other higher education systems offer programs of study in Food Science, Food Technology, Food Business and Marketing and/or Food Production. Of course, we all must eat and with a large portion of us eating out these days, there are many opportunities in the restaruant, hospitality, hotel and hospital trade for food trained personnl to excel. Food is essential, but in most cases it must be preserved to make it available and the food processing business is the connecting link in that great chain known as "from the gate to the plate".

Chapter 2

UNDERSTANDING OUR PAST

History tells us where we have been and gives us insight for the future that could lead us to where we want to go.

Or, as Calvin Coolidge states, *"No person was ever honored for what he received. Honor has been the reward for what he gave."*

No industry just happened. It takes vision, experimenting and testing and implementing plans to assure that they work. The food preservation industry is no exception. Many people have spent their lifetime making things better for us all. Certainly the food industry, with the help of the Good Lord, has come a long way since the early beginning as recorded in the Bible where man first learned to dry foods, ferment foods, bake, and, yes, even keep them cold.

However, it wasn't until 1795 when the armies of England, Prussia, Austria, Spain and Sardinia were fighting France that food became the number one concern. Napoleon found out that an army can move only as fast as it can provide food for the soldiers to eat. Foraging was inefficient and certainly most inconvenient. Even though the French army won many battles, they lost many lives due to scurvy and other forms of malnutrition. It is reported that the French Directory offered 12,000 francs to the patriot who could devise a new and successful method of preserving food for long periods.

Nicholas Appert, an obscure citizen who had been a chef, pickler, preserver, winemaker, brewer, confectioner, and distiller, entered upon a fourteen year period of experimentation on food preservation. In 1810 he won the coverted prize. His studies were published in a book entitled, *The Art of Preserving All Animal and Vegetable Substances for Several Years.* Appert demonstrated that heat applied to food sealed in an airtight container kept food from spoiling.

Appert used wide mouth glass bottles which he filled with food, corked them and then heated the bottles and foods in boiling water. His published works revealed that he had successfully preserved more than 50 kinds of food. He didn't know why they kept, but many of his recommendations are still followed today, such as, complete cleanliness, absolute necessity of sealing the container to exclude air, and a thorough cooking of the food in the sealed container.

In 1810, Peter Durand obtained a patent covering the use of iron and tin to make canisters for use in preserving food. These were the forerunners to what we know today as the tin can or "can."

In 1820, Thomas Kensett in New York and William Underwood in Boston began putting up foods by the Appert method using glass jars sealed with tightly wired cork stoppers. However, the glass container industry was slow in commercializing as the fruit type jar was not readily available until the 1870's. Kensett and Underwood in 1839 began packing their products in metal containers similar to Durand's. The sales people shortened the canister (from the Greek word *kanastron*, to "can").

It should be noted that Louis Pasteur in 1860 supplied the missing link as to why Appert's foods kept from spoiling. He is given credit for being the father of food microbiolgy. He showed that microorganisms can be destroyed by heat, the major principle in understanding canning.

Perhaps at this point we should define what is meant by the term "canning." Canning is a method of food preservation wherein a food and its container are rendered commercially sterile by the application of heat, alone or in combination with given pH and/ or water activity control. Commercially sterile simply means destruction of all public health organisms capable of growing in the food under normal nonrefrigerated conditions of storage and distribution.

Before the turn of the century, these so-called "cans" were all hand made from imported tin-plated sheet iron. The bodies were cut with shears, rolled into cylinders and the side seam was hand soldered. The bottom and top ends were hand soldered in place with a hole left in the top end of the can. The food was stuffed into the can through the hole and when filled, the hole in the top of the can was covered with a tin/metal disk and soldered closed. The can, with its contents of food, was then cooked or processed. A good can-maker could make some "hole-in-top" cans at the rate of 60 cans in a 10-hour day.

According to records on file by the Mid-America Food Processors Association, the modern so-called "sanitary can" was unveiled in Columbus, Ohio in the early 1900's. This container had one end soldered in place, but the other end was mechanically applied using a crimping and sealing machine—known today as the double-seaming machines. The records show the industry did not start to use this container until 1916 because they did not believe it would work. At a meeting in the Southern Hotel, Columbus, Ohio, the safety of the seal was accepted. This new container was called the "sanitary can." The sanitary can was so labeled because the can could be washed and drained prior to filling.

Today, the sanitary can is made from steel sheets that are coated with tin at the rate 0.25 to 2% tin. The tin and/or other coatings are necessary to keep the food from reacting with the steel.

Aluminum cans appeared on the market in the 1960's. They were used in the beer and beverage industry because they were very light in weight, thus cutting shipping costs. The cans could be cooled quickly because of the thermal conductivity of aluminum and they are reported to provide the contents with a longer shelf life than beverage or food packed in tin cans. The aluminum cans were the first cans to be made by stamping them out. A similar process, that is, the "drawing and ironing" of a steel glob, is used today for many metal cans. These types of two-piece containers are seamless and may have lids or a top that can be pulled open without using the conventional can opener.

Plastic containers developed in the late 1980's are finding some use in food preservation, particularly in the aseptic processing of foods. Their advantage is that the plastic container may be made in house and the food can be microwaved right in the container for reheating prior to consumption.

Even though the package is the most important aspect of preserving heat and other methods of processed foods, many other contributions were being made by the other suppliers.

The first of these is Isaac Soloman, a Baltimore canner who adopted a method of raising the boiling point of water developed by Sir Humphry Davy, an English chemist, by adding calcium chloride to the water to raise the boiling point from 212°F (100°C) to 240°F (115.6°C) Solomon reduced the time to sterilize his food products from 5–6 hours to 30 minutes.

A. K. Shriver in 1874 developed a steam pressure kettle or as we know it today, the retort or pressure cooker. This was a major step forward in controlling the cook time and preserving the quality of the product.

During the period between 1894 and 1905, H. L. Russell, U. S. Department of Agriculture in Wisconsin, Dr. Samuel Cate Prescott of MIT, and William L. Underwood, grandson of the pioneer Boston canner, reported on the reasons that cans of food blow up. Their presentation to a canning industry meeting in 1905 in Cleveland, Ohio showed that insufficient heat caused the food in the cans to develop gas from spoilage microorganisms. The gas caused the cans to explode. These studies were the start of scientific heat requirements to be used when processing canned foods, but the true application of heat requirements was not known until C. Olin Ball in 1918 developed his famous mathematical data for heat processing. He illustrated quite clearly how heat penetrates into the food in the container and the need to understand given center temperatures depending on the product(s).

In 1907 the National Canners Association was formed with headquarters in Washington, D. C. This association established food research and a service laboratory to help the industry help themselves. The can and glass manufacturing firms, also, established laboratories to help their customers better understand food preservation.

Starting in the early 1900's, workers at the U. S. Department of Agriculture contributed to the science of food preservation, with A. W. and K. G. Bitting, C. A. Magoon and C. W. Culpepper, conducting blanching and processing studies.

In the 1920's, personnel at MIT, University of Massachusetts, University of California, Oregon State University, University of Wisconsin and other universities established courses of instruction to train personnel for the food industry. Along with the training at these universities, faculty initiated investigations on food preservation and the technology of production of crops specifically for the food processing industry.

Much research work was being conducted by supply people, canners, and by the faculty in some of the leading universities all working together with industry to better understand microbes, containers, processes, and food product quality. I have listed some of these early leaders and their contributions in Tables 2.1 and 2.2.

There are many more persons who should receive recognition for their many contributions, but this list will give one a little more knowledge about the early leaders. The references attached will provide further reading for those interested.

TABLE 2.1 — Some Personnel in the Early Development
of the Food Processing Industry

Date	Person(s)	Contribution
1809	Nicolas Appert	Father of Canning
1810	Peter Durand	Development of tin can
1839	Isaac Winslow	1st U.S. canner
1860	Louis Pasteur	Father of Microbiology
	Isaac Soloman	Salt used to raise Boiling Point of water
1873	Andrew Shriver	Development of Retort
1885	Welcome Sprague	Developed Corn Cutter
1890	Max Ams	Developed Can Closing Machine
1894	Harry Russell	Microbiologist (U of WI) studied cause of can spoilage
1898	Wm. Underwood	Canner working with Prescott
1898	Sam C. Prescott	Professor at MIT working with Russell and Underwood on food spoilage
1903	Harvey W. Wiley	USDA chemist,"Father of Food Laws"
1916	A.W. & K.G. Bitting	USDA food chemists developed procedures for canning
1920	C. Olin Ball	Developed Mathematical Processing Tech.
1921	C.A. Magoon & C.W. Culpepper	Technology of Blanching know-how
1923	Clarence Birdseye	Father of Frozen Foods

TABLE 2.2 — Nicholas Appert IFT Awardees

(Preeminence in and contributions to the field of food technology)

Date	Name	Affiliation
1942	William V. Cruess	University of California
1943	Samuel C. Prescott	Massachusetts Institute of Technology
1944	C. A. Browne	U. S. Department of Agriculture
1945	A. W. Bitting	NCA & Glass Container
1946	C. H. Bailey	University of Minnesota
1947	C. O. Ball	NCA, Am. Can., OI Glass
1948	C. A. Elvehjem	University of Wisconsin
1949	Roy C. Newton	Purdue University
1950	Thomas M. Rector	General Mills
1951	A. E. Stevenson	Continental Can Co.
1952	Edward M. Chase	U. S. Department of Agriculture
1953	Victor Conquest	Armour & Co.
1954	Charles N. Frey	Fleishman Labs.
1955	Charles G. King	Nutrition Foundation
1956	Bernard E. Proctor	Massachusetts Institute of Technology
1957	Emil M. Mrak	University of California
1958	William F. Geddes	University of Minnesota
1959	Berton S. Clark	American Can Co.
1960	Ernest H. Weigand	University of Oregon
1961	Helmut C. Diehl	Refrigeration Foundation
1962	Arnold Kent Balls	
1963	Karl F. Myers	University of California
1964	Gail M. Dack	
1965	Harold W. Schultz	University of Oregon
1966	Maynard A. Joslyn	University of California
1967	Michael J. Copley	
1968	Donald K. Tressler	AVI Publishing Company
1969	Edwin M. Foster	
1970	Samuel A. Golblith	Massachusetts Institute of Technology
1971	Reid T. Milner	University of Illinois
1972	John C. Ayers	University of Georgia
1973	Hans Lineweaver	U. S. Department of Agriculture
1974	George F. Stewart	University of California
1975	Ernest J. Briskey	Armour
1976	Amihud Kramer	University of Maryland
1977	Richard L. Hall	McCormick & Co.
1978	Jasper G. Woodruff	University of Georgia

(Table continues next page)

TABLE 2.2 — Nicholas Appert IFT Awardees -(Continued)

Date	Name	Affiliation
1979	Frederick J. Francis	Univeristy of Massachusetts
1980	Evan C. Binkerd	
1981	B. S. Schweigert	University of California
1982	C. O. Chichester	
1983	Stephen S. Chang	Rutgers University
1984	John J. Powers	Univeristy of Georgia
1985	Alina Szczesniak	
1986	Marcul Karel	
1987	Elmer H. Marth	
1988	Owen R. Fenema	University of Wisconsin
1989	Fergus Clydesdale	University of Massachusetts
1990	Myron Solberg	
1991	Raymond J. Moshy	Hunt/Wesson Foods
1992	Irving J. Pflug	University of Minnesota
1993	Wilbur A. Gould	Ohio State University
1994	Roy L. Whistler	
1995	Phillip E. Nelson	Purdue University
1996	Michael P. Doyle	University Of Georgia
1997	H. C. Rudlph Heiss	Institute Of Food Technology & Packaging, Munich, Germany

TABLE 2.3 — Some Notable Machinery and New Preserved
Products During Recent Decades

Decade	Machine or Services	Product
< 1860	All Hand Operations	Tomatoes Peas Cream Style Corn Condensed Milk Peaches Cranberries Lobsters & Oysters
1860–1870	Corn Cooker Calcium Chloride Kettle	Asparagus, white Apricots Blueberries Catsup Cherries Corned Beef Pineapple Pears and Plums Soup Salmon & Sardines
1870–1880	Steam Pressure Cookers	Apples Crab Meat Lima Beans Macaroni Pumpkin & Squash Sweet Potato Shrimp Pork and Beans
1880–1890	Automatic Fillers Automatic Labelers Cream Style Corn Cutter Pea Huller & Size Grader Corn Huskers & Silkers Can Lacquers	Evaporated Milk Sauerkraut
1890–1900	Pea Blancher Pea Viner & Gravity Seapartor Tomato Scalder & Cyclone Apple Peeler	

(Table continues next page)

TABLE 2.3 — Some Notable Machinery and New Preserved
Products During Recent Decades - (Continued)

Decade	Machine or Services	Product
1900–1910	Lye Peeler Sanitary Can Bean Cutter, Snipper & Grader Abrasive Peeler Cherry Pitter	Whole Grain Corn Olives Mushrooms Spinach Tuna and Figs
1910–1920	Ginaca Machine for Pineapples Corn Cutter for Whole Kernel Peach Pitter	Pineapples Artichoke Grapefruit Green Asparagus Dried Prunes Pimentoes Cider Ketchup
1920–1930	Juice Extractors Tubular Blanchers Stainless Steel Kettles	Tomato Juice Strained Foods Pineapple Juice Cranberry Sauce Dog Food
1930–1940	Heat Unit System/Crop Maturity Pear Peeler & Corer Steam Pressure Peeler Homogenizer Red Beet & Carrot Harvesters Vacuum Concentrators	Orange Juice Fruit Cocktail Whole Ham and Fowl Potato Chips
1940–1950	Aseptic Canning for Juices Palletizers Electric Eye Sorters Corn & Spinach Harvesters	Corn Chips Pie & Pastry Fillings Orange Concentrate
1950–1960	Snap Bean Harvesters Cemented Side Seam Cans Tomato Harvesters Color Sorters Nitrogen & Carbon Dioxide Freezers Hydrostatic Cookers CA Storages Reverse Osmosis	Soft Drinks Tostidos Frozen TV Dinners Diced Tomatoes

(Table continues next page)

TABLE 2.3 — Some Notable Machinery and New Preserved
Products During Recent Decades - (Continued)

Decade	Machine or Services	Product
1960–1970	Pickle Harvesters Electric Welded Cans Aluminum Cans Two-Piece Metal Cans Ultrafiltration UHT Processing of Fluid Products	Aseptic Puddings Fruit Drinks Kiwi Fruit Salsas
1970–1980	Iron & Drawn Cans Computers Automation Robotics CIP Systems Waste Treatments & Utilization	
1980–1990	Microwave Cooking	Biotechnoogy Genetic Engineered Tomatoe Potatoes Corn
1990s	Pulse Electric Heating Ohmic Heating	

TABLE 2.3 — Some Significant Dates & Activities In U. S. Food
Law History

Year(s)	Activity
1784	Massachusetts enacted the first general food law in U.S.
1850	California passed a pure food and drink law
1879–1905	During this 26 years, more than 100 food and drug bills were introduced in Congress
1879	Peter Collier, chief chemist, USDA, began investigating food adulteration
1883	Dr. Harvey W. Wiley became chief chemist, USDA, on April 9, and assigned members of his staff to study the problems of food adulteration
1897	The Tea Importation Act was passed providing for inspection of all tea entering the U.S.
1902	Congress made appropriations to establish pure food standards
1906	The Pure Food and Drugs Act passed by Congress and signed into law by President Theodore Roosevelt
1907	The Bureau of Chemistry began administering the new Food and Drug Act
1913	Gould Amendment enacted, requiring that definite quantity information appear on food packages
1930	McNary–Mapes Amendment enacted, authorizing Standards of Quality and Fill of Container for canned food
1938	The Federal, Food, Drug and Cosmetic Act enacted, authorizing Standards of Identity, Quality and Fill of Container for foods, and authorizing factory inspections
1939	First food standards issued July 14 for canned whole tomatoes, tomato puree and tomato paste
1940	FDA transferred from USDA to Federal Security Agency with Walter G. Campbell as first Commissioner of Food and Drugs
1951	Supreme Court ruled that a standardized food which does not meet its Standard of Identity cannot be sold unless it is marked "imitation"
1952	Supreme Court ruled the factory inspection provision of FDC Act was too vague to be enforced as criminal law (U.S. vs. Cardiff)
1953	Federal Security Agency became Department of Health, Education and Welfare (DHEW)

(Continues next page)

1953 (cont.)	Factory Inspection Amendment clarified previous law and required FDA to give manufacturers written reports of inspection and analysis of factory samples
1954	Hale Amendment simplified method of promulgating food standards when no controversy was involved
	Miller Pesticide Amendment streamlined procedures for setting safety limits for pesticide residues on raw agricultural commodities and greatly strengthened consumer protection
1958	Food Additive Amendment enacted, prohibiting use of new food additives until sponsor establishes safety and FDA issues regulations specifying conditions of use
1960	Color Additive Amendment enacted to allow FDA to establish by regulations the conditions of safe use for color additives in foods, and to require manufacturers to perform necessary scientific investigations to establish safety
1966	Fair Packaging and Labeling Act enacted to require that consumer products in interstate commerce be honestly and informatively labeled
1969	Regulation, Part 110 promulgated dealing with Current Good Manufacturing Practice (CGMP) in manufacturing, packing or holding human food
1973	Regulation, Part 113 promulgated dealing with (Modified Thermally Processed Low-Acid Foods packaged in 1978) hermetically sealed containers
1978	Part 109–Recall Guidelines promulgated
1979	Regulation, Part 114 promulgated dealing with Acidified Foods
1986	Part 318 of 9 CFR–Meat & Poultry Canning Regulations
1990	Nutritional Labeling and Education Act enacted
1995	Regulation, Part 123 promulgated dealing with Procedures for the Safe and Sanitary Processing and importing of fish and fishery products mandating application of Hazard Analysis Critical Control Point (HACCP) principles to the processing of seafood
1996	USDA, Food Safety and Inspection Service published *Pathogen Reduction; Hazard Analysis Critical Control Point (HACCP) Systems* — Final Rule, July 1996
	H.R. 1627, The Food Quality Protection Act...Delaney Clause zero risk standard for pesticide residues eliminated and replaced with a unified standard of "safe" for processed and raw foods

Chapter 3

THE SIGNIFICANCE OF MICROORGANISMS

"What Doesn't Kill You Will Make You Stronger"
Chinese Proverb

Microorganisms are practically everywhere, on everything, invisible and maybe useful or harmful to mankind, plants, animals and the very food that we eat. They include viruses, bacteria, protozoa, algae, fungi, and certain small worms. There are some organisms that may present serious public health problems in terms of food safety, food spoilage, and/or food preservation. Some of these microorganisms have caused serious outbreaks in recent years.

On the other hand, many organisms are valuable to food production and food preservation. These organisms are responsible for the production of many crops, the manufacture of sauerkraut from cabbage, pickles from cucumbers, and cheese from milk. Other microorganisms are responsible for many medicines, organic acid production, and enzymes.

Microorganisms may be classified according to their optimal temperature for growth, peculiar growth requirements (sugars and salt), and or their major nutrient source. **Psychrotropic** organisms grow at relatively low temperatures including commercial refrigerator temperatures, **thermophilic** organisms survive and grow well at elevated temperatures, while **mesophilic** organisms are both aerobic and anaerobic sporeforming bacteria that can grow at 35°C but not at 55°C.

All species of the genus Bacillus are aerobic. Anaerobic spore forming mesophillic organisms that cause food spoilage by putrefaction include *Clostridium sporogenes, C. botulinum, C. perfringens* and *C. putrefaciens*. Most mesophilic organisms are heat resistant.

Proteolytic microorganism break down proteins, **lipolytic**

microorganisms breakdown lipids or fats, **Saccharolytic** or acid forming microorganisms breakdown sugars and acids, **pectinolytic** microorganisms break down pectins and are the cause of onset of spoilage in many fruits and vegetables, **amylolytic** microorganisms break down starch to sugars, and **cellulytic** microorganisms break down cellulose to simpler carbohydrate compounds.

Microorganisms may, also, be helpful in predicting if pathogenic organisms may be present. These are called indicator organisms, such as *Escherichia coli*. The presence or absence of bacterial toxins is also used as an indicator for pathogens and as a means to help insure the safety of the food.

Microorganisms differ in their size and shape as well as their biochemical and cultural characteristics. They cannot be seen by the naked eye, but under a microscope one can discern their shape and, in some cases, their characteristics such as spore forming ability.

MOLDS

Molds are without doubt an indicator organisms in terms of food plant sanitation. Molds exhibit some of the characteristics of higher plants. There are several distinguishing characteristics of Molds, such as, (1) multicellular, (2) vegetative filaments called mycelia and hyphae which are usually branched and round and always parallel, (3) mold filaments which are granular or stippled in appearance, that is, they appear normally filled with food, (4) filaments with cross walls or septate in appearance, and (5) ends which are usually blunt and hardly ever pointed.

Molds may be classified on basis of function, that is the vegetative cells versus the fertile cells. The vegetative filaments are those that grow under the surface of substances while the fertile filaments are those which bear the fruiting bodies or reproductive spores and they grow on the surface of food, equipment or even walls, ceilings and portions of buildings. The combination of these two form the entire mold plant with the submerged filaments serving to anchor the mold while the above portion serving to provide air to the mold plant.

Molds are much larger than bacteria and somewhat larger than yeasts. Molds are widely distributed in nature, found in air and in the soil as well as on plants. Molds will grow under wide latitudes of moisture, air, and temperature.

Molds may be useful to mankind in that they provide interesting flavors to many cheeses and they are useful in certain enzyme production. Molds may contribute to spoilage and some produce undesirable toxins.

Molds can be distinguished, in part, by the color of their fruiting bodies, that is, black molds (*Aspergillus*), green molds (*Penicillium*), white molds (*Geotrichum*), and pink molds (*Neurospora*).

YEAST

Yeast are, also, plant like substances except they have no chlorophyll and they are dependent on plants or animals for their energy. Thus, they are considered a parasite or a saprasite. Yeasts are unicellular and they may be cylindrical or elongated and sometimes ellipsoidal. Yeasts range widely in size, but are smaller than molds and most yeast are larger than bacteria. Yeasts may reproduce asexually by budding or they may reproduce sexually. Yeasts produce enzymes that effect many desirable reactions in foods, such as, leavening of bread and the production of alcohol and invert sugar. Like molds, yeasts are more tolerant to cold than heat with most yeast forms destroyed at temperatures of approximately 180°F (27.8°C). Yeasts generally do not present a public health problem.

BACTERIA

Bacteria are very abundant in nature and they can be very difficult in terms of the enzymes they excrete which may produce undesirable changes in food. Bacteria are single celled living bodies so small that they can only be seen with a microscope. They vary in length from $1/25,000$ to $1/1,000$ of an inch.

Bacteria have several shapes from round (cocci) to rods. Bacteria reproduce by dividing, called fision. In many cases they can divide every 20 minutes. This division is referred to as the growth rate. Thus, at the end of one hour, one bacteria will produce 4 new bacteria and at the end of 2 hours there are 16 new bacteria or at the end of 15 hours some 1,000,000,000 bacteria.

Bacteria must have food to grow and man can control their growth by limiting food or providing food of different pHs. If the conditions are not favorable for growth, some bacteria may produce spores. All of the round bacteria and many of the rod shaped bacteria cannot produce spores, thus they are referred to as non-sporeformers.

Sporeformers, on the other hand, are generally very resistant to heat, cold, and many chemicals. For example, they can survive temperatures above 212 degree F. (100 degree C.) for many hours.

Probably the best known type of bacteria is *Clostridium botulinum*. This organism is of major concern to the food industry, particularly the canning industry. If this spore-forming, rod- shaped organism is not destroyed during heating, it may produce a deadly toxin or poison. All non-acid (low, that is, the pH in excess of 4.6) type foods requires a pressure process time and temperature sufficient to destroy the spores of this organism. The organism is found everywhere and the vegetative cells of *C. botulinum* can grow in the absence of air or oxygen (anaerobic conditions) and produce the toxin. However, the toxin, should it be produced, can be inactivated by boiling temperatures. Further, *C. botulinum* will not grow in acid type foods, thus acidic foods need not be pressure processed as only vegetative cells need be destroyed.

For microorganisms to grow and/or survive they need water and in recent years food microbiologists tell us that the amount of free water has a direct bearing on microbial growth. *C. botulinum* will not grow if the water activity (a measure of free water in the food) level is below 0.85.

Control of microorganisms is a major requirement in every food processing operation. It all starts with the raw material and its method of preparation, that is, washing, sorting, blanching, etc., and, most importantly, the maintenance of proper food plant sanitation. Food plant sanitation includes all the equipment, the premises, the packaging, the handling of the food and its proper processing (heat cold, or chemical methods) and the people working in the fields, factories, warehouses, etc. Many people will agree that people may cause food poisoning outbreaks because of their lack of knowledge of the food and the microorganisms that may be present. The following information may be most helpful in understanding the many common foodborne diseases caused by bacteria, viruses, fungi, protozoa and parasites, chemicals and metals including the symptoms, particular foods, mode of contamination and how to prevent the particular diseases.

TABLE 3.1 — Some Characteristics for Growth
of Specific Microorganisms

Pathogen	Temperature	pH	Minimum Water Activity (A_W)
Molds	34–110°F 1.1–43°C	3-6	0.75
Yeasts	34–110°F 1.1–43°C	3-6	0.88
C. botulinum (Types A, B, E)	38–115°F 3.3–46°C	>4.6	0.94
C. perfringens	59–122°F 15–50°C	5.5–8.0	0.95
E. coli 0157:H7	50–108°F 10–42°C	4.5–9.0	——
Salmonella	41–116°F 5–46°C	4–9	0.94
Listeria monocytogenes	36–115°F 2.5–44°C	5.2–9.6	——
Bacillus cereus	50–120°F 10–48°C	4.9–9.3	0.95
Staphylococcus aureus	44–115°F 6.5–46°C	5.2–9.0	0.86

TABLE 3.2 — Common Foodborne Diseases Caused by Bacteria

Disease (causative agent)	Latency Period (duration)	Principal Symptoms	Typical Foods	Mode of Contamination	Prevention of Disease
(Bacillus cereus) food poisoning, diarrheal	8–16 hours (12–24 hrs.)	Diarrhea, cramps, occasional vomiting	Meat products, soups, sauces, vegetables	From soil or dust	Thorough heating and rapid cooling of foods
(Bacillus cereus) food poisoning, emetic	1–5 hours (6–24 hrs.)	Nausea, vomiting, sometimes diarrhea & cramps	Cooked rice and pasta	From soil or dust	Thorough heating and rapid cooling of foods
Botulism; food poisoning (heat-labile toxin of Clostridium botulinum)	12–36 hours (months)	Fatigue, weakness, double vision, slurred speech, respiratory failure, sometimes death	Types A&B: vegetables, fruits, meats, fish and poultry products, condiments; Type E: fish and fish products	Types A&B: from soil or dust; Type E: water and sediments	Thorough heating and rapid cooling of foods
Botulism; food poisoning; infant infection	Unknown	Constipation, weakness, respiratory failure, sometimes death	Honey, soil	Ingested spores from soil or dust or honey colonize intestine	Do not feed honey to infants — will not prevent all
Campylobacteriosis (Campylobacter jejuni)	3–5 days (2–10 days)	Diarrhea, abdominal pain, fever, nausea, vomiting	Infected food – source animals	Chicken, raw milk	Cook chicken thoroughly; avoid cross-contamination; irradiate chickens; pasteurize milk
Cholera (Vibrio cholerae)	2–3 days (hours to days)	Profuse watery stools, sometimes vomiting, dehydration, often fatal if untreated	Raw or undercooked seafood	Human feces in marine environment	Cook seafood thoroughly; general sanitation
(Clostridium perfringens) food poisoning	8–22 hours (12–24 hrs.)	Diarrhea, cramps, rarely nausea and vomiting	Cooked meat and poultry	Soil, raw foods	Thorough heating and rapid cooling of foods
(Escherichia coli) foodborne infection: enterohemorrhagic	12–60 hours (2–9 days)	Watery, bloody diarrhea	Raw or undercooked beef, raw milk	Infected cattle	Cook beef thoroughly; pasteurize milk

[Continued]

TABLE 3.2 — Common Foodborne Diseases Caused by Bacteria - (Continued)

Disease (causative agent)	Latency Period (duration)	Principal Symptoms	Typical Foods	Mode of Contamination	Prevention of Disease
(*Escherichia coli*) foodborne infection: enteroinvasive	At least 18 hours (uncertain)	Cramps, diarrhea, fever, dysentery	Raw foods	Human fecal contamination, direct or via water	Cook foods thoroughly; general sanitation
(*Escherichia coli*) foodborne infection: enterotoxigenic	10–72 hours (3–5 days)	Profuse watery diarrhea, sometimes cramps and vomiting	Raw foods	Human fecal contamination, direct or via water	Cook foods thoroughly; general sanitation
Listeriosis (*Listeria monocytogenes*)	3–70 days	Meningoencephalitis; stillbirths; septicemia or meningitis in newborns	Raw milk, cheese and vegetables	Soil or infected animals, directly or via manure	Pasteurization of milk; cooking
Salmonellosis (*Salmonella* species)	5–72 hours (1–4 days)	Diarrhea, abdominal pain, chills, fever, vomiting, dehydration	Raw and undercooked eggs; raw milk, meat and poultry	Infected food – source animals; human feces	Cook eggs, meat & poultry thoroughly; pasteurize milk; irradiate chickens
Shigellosis (*Shigella* species)	12–96 hours (4–7 days)	Diarrhea, fever, nausea; sometimes vomiting, cramps	Raw foods	Human fecal contamination, direct or via water	General sanitation; cook foods thoroughly
Staphylococcal food poisoning (heat-stable enterotoxin of *Staphylococcus aureus*)	1–6 hours (6–24 hrs.)	Nausea, vomiting, diarrhea, cramps	Ham, meat and poultry products, cream-filled pastries, whipped butter, cheese	Handlers with colds, sore throats or infected cuts, food slicers	Thorough heating and rapid cooling of foods
Streptococcal foodborne infection (*Streptococcus pyogenes*)	1–3 days (varies)	Various, including sore throat, erysipelas, scarlet fever	Raw milk, deviled eggs	Handlers with sore throats, other "strep" infections	General sanitation; pasteurize milk

[Continued]

TABLE 3.2 — Common Foodborne Diseases Caused by Bacteria - (Continued)

Disease (causative agent)	Latency Period (duration)	Principal Symptoms	Typical Foods	Mode of Contamination	Prevention of Disease
Vibrio parahaemolyticus foodborne infection	12–24 hours (4–7 days)	Diarrhea, cramps; sometimes nausea, vomiting, fever, headache	Fish and seafoods	Marine coastal environment	Cook fish and seafoods thoroughly
Vibrio vulnificus foodborne infection	In persons with high serum iron: 1 day	Chills, fever, prostration, often death	Raw oysters and clams	Marine coastal environment	Cook shellfish thoroughly
Yersiniosis (*Yersinia enterocolitica*)	3–7 days (2–3 weeks)	Diarrhea, pains mimicking appendicitis, fever, vomiting, etc.	Raw or undercooked pork and beef; tofu packed in spring water	Infected animals, especially swine; contaminated water	Cook meats thoroughly; chlorinate water

TABLE 3.3 — Common Foodborne Diseases Caused by Viruses

Disease (causative agent)	Latency Period (duration)	Principal Symptoms	Typical Foods	Mode of Contamination	Prevention of Disease
Hepatitis A (Hepatitis A virus)	15–50 days (weeks to months)	Fever, weakness, nausea, discomfort; often jaundice	Raw or undercooked shellfish; sandwiches, salads, etc.	Human fecal contamination, via water or direct	Cook shellfish thoroughly; general sanitation
Viral gastroenteritis (Norwalk-like viruses)	1–2 days (1–2 days)	Nausea, vomiting, diarrhea, pains, headache, mild fever	Raw or undercooked shellfish; sandwiches, salads, etc.	Human fecal contamination, via water or direct	Cook shellfish thoroughly; general sanitation
Viral gastroenteritis (rotaviruses)	1–3 days (4–6 days)	Diarrhea, especially in infants and young children	Raw or mishandled foods	Probably human fecal contamination	General sanitation

TABLE 3.4 — Common Foodborne Diseases Caused by Fungi Other Than Mushrooms

Disease (causative agent)	Latency Period (duration)	Principal Symptoms	Typical Foods	Mode of Contamination	Prevention of Disease
Aflatoxicosis ("aflatoxins" of *Aspergillus flavus* and related molds)	Varies with dose	Vomiting, abdominal pain, liver damage; liver cancer (mostly Africa and Asia)	Grains, peanuts, milk	Molds grow on grains and peanuts in field or storage; cows fed moldy grain	Prevent mold growth; don't eat or feed moldy grain or peanuts; treat grain to destroy toxins
Alimentary toxic aleukia ("trichothecene" toxin of *Fusarium* molds)	1-3 days (weeks to months)	Diarrhea, nausea, vomiting; destruction of skin & bone marrow; sometimes death	Grains	Mold grows on grain, especially if left in field through winter	Harvest grain in fall; don't use moldy grain
Ergotism (toxins of *Claviceps purpurea*)	Varies with dose	Gangrene (limbs die and drop off); or convulsions and dementia; abortion (now not seen in U.S.)	Rye; or wheat, barley and oats	Fungus grows on grain in field; grain kernel is replaced by a "sclerotium"	Remove sclerotia from harvested grain

TABLE 3.5 — Common Foodborne Diseases Caused by Protozoa & Parasites

Disease (causative agent)	Latency Period (duration)	Principal Symptoms	Typical Foods	Mode of Contamination	Prevention of Disease
(PROTOZOA) Amebic dysentery (*Entamoeba histolytica*)	2–4 weeks (varies)	Dysentery, fever, chills; sometimes liver abscess	Raw or mishandled foods	Cysts in human feces	General sanitation; thorough cooking
Cryptosporidiosis (*Cryptosporidium parvum*)	1–12 days (1–30 days)	Diarrhea; sometimes fever, nausea and vomiting	Mishandled foods	Oocysts in human feces	General sanitation; thorough cooking
Giardiasis (*Giardia lamblia*)	5–25 days (varies)	Diarrhea with greasy stools, cramps, bloat	Mishandled foods	Cysts in human and animal feces; directly or via water	General sanitation; thorough cooking
Toxoplasmosis (*Toxoplasma gondii*)	10–23 days (varies)	Resembles mononucleosis; fetal abnormality or death	Raw or undercooked meats; raw meat; mishandled foods	Cysts in pork or mutton, rarely beef; oocysts in cat feces	Cook meat thoroughly; pasteurize milk; general sanitation
(ROUNDWORMS, Nematodes) Anisakiasis *Anisakis simplex*, *Pseudoterranova decipiens*)	Hours to weeks (varies)	Abdominal cramps, nausea, vomiting	Raw or undercooked marine fish, squid or octopus	Larvae occur naturally in edible parts of seafoods	Cook fish thoroughly or freeze at –4°F for 30 days

[Continued]

TABLE 3.5 — Common Foodborne Diseases Caused by Protozoa & Parasites - (Continued)

Disease (causative agent)	Latency Period (duration)	Principal Symptoms	Typical Foods	Mode of Contamination	Prevention of Disease
Ascariasis (*Ascaris lumbriocoides*)	10 days–8 weeks (1–2 years)	Sometimes pneumonitis, bowel obstructions	Raw fruits or vegetables that grow in or near soil	Eggs in soil from human feces	Sanitatary disposal of feces; cooking food
Trichinosis (*Trichinella spiralis*)	8–15 days (weeks, months)	Muscle pain, swollen eyelids, fever; sometimes death	Raw or undercooked pork or meat of carnivorous animals (e.g., bears)	Larvae encysted in animal's muscles	Thorough cooking of meat; freezing pork at 5°F for 30 days; irradiation
(TAPEWORMS, Cestodes) Beef tapeworm (*Taenia saginata*)	10–14 weeks (20–30 years)	Worm segments in stool; sometimes digestive disturbances	Raw or undercooked beef	"Cysticerci" in beef muscle	Cook beef thoroughly or freeze below 23°F
Fish tapeworm (*Diphyllobothrium latum*)	3–6 weeks (years)	Limited; sometimes vitamin B-12 deficiency	Raw or undercooked freshwater fish	"Plerocercoids" in fish muscle	Heat fish 5 minutes at 133°F or freeze 24 hours at 0°F
Pork tapeworm (*Taenia solium*)	8 weeks–10 years (20–30 years)	Worm segments in stool; sometimes "cysticercosis" of muscles, organs, heart or brain	Raw or undercooked pork; any food mishandled by a *T. solium* carrier	"Cysticerci" in pork muscle; any food handled by human feces with *T. solium* eggs	Cook pork thoroughly or freeze below 23°F; general sanitation

TABLE 3.6 — Common Foodborne Diseases Caused by Chemicals & Metals

Disease (causative agent)	Latency Period (duration)	Principal Symptoms	Typical Foods	Mode of Contamination	Prevention of Disease
(TOXINS IN FINFISH) Ciguatera poisoning (ciguatoxin, etc.)	3-4 hours (rapid onset)	Diarrhea, nausea, vomiting, abdominal pain	"Reef and island" fish: grouper, surgeon fish, barracuda, pompano, snapper, etc.	(Sporadic); food chain, from algae	Eat only small fish
	12–18 hours (days–months)	Numbness and tingling of face; taste and vision aberrations; sometimes convulsions; respiratory arrest and death (1-24 hrs)			
Fugu or pufferfish poisoning (tetrodotoxin, etc.)	10–45 min. to ≥ 3 hrs.	Nausea, vomiting, tingling lips & tongue, ataxia, dizziness, respiratory distress/arrest, sometimes death	Pufferfish, "fugu" (many species)	Toxin collects in gonads, viscera	Avoid pufferfish (or their gonads)
Scombroid or histamine poisoning (histamine, etc.)	Minutes to few hours (few hours)	Nausea, vomiting, diarrhea, cramps, flushing, headache, burning in mouth	"Scombroid" fish (tuna, mackerel, etc.); mahimahi, others	Bacterial action	Refrigerate fish immediately when caught
(TOXINS IN SHELLFISH) Amnesic shellfish poisoning (domoic acid)		Vomiting, abdominal cramps, diarrhea, disorientation, memory loss; occas. death	Mussels, clams	From algae	Heed surveillance warnings
Paralytic shellfish poisoning (saxitoxin, etc.)	≤ 1 hour (≤ 24 hrs)	Vomiting, diarrhea, paresthesias of face, sensory & motor disorders; respiratory paralysis, death	Mussels, clams, scallops, oysters	From "red tide" algae	Heed surveillance warnings
(MUSHROOM TOXINS) Mushroom poisoning (varies greatly among species)	< 2 hrs. to ≥ 3 days	Nausea, vomiting, diarrhea, profuse sweating, intense thirst, hallucinations, coma, death	Poisonous mushrooms	Intrinsic	Don't eat wild mushrooms

[Continued]

TABLE 3.6 — Common Foodborne Diseases Caused by Chemicals & Metals - (Continued)

Disease (causative agent)	Latency Period (duration)	Principal Symptoms	Typical Foods	Mode of Contamination	Prevention of Disease
(PLANT TOXINS) Cyanide poisoning (cyanogenetic glycosides from plants)	(Large doses) 1–15 minutes	Unconsciousness, convulsions, death	Bitter almonds, cassava, some lima bean varieties, apricot kernels	Intrinsic, natural	Proper processing; avoid some so-called foods
(METALS) Cadmium	Depends on dose	Nausea, vomiting, diarrhea, headache, muscular aches, salivation, abdominal pain, shock, liver damage, renal failure	Acid foods, foods grilled on shelves from refrigerator	Acid or heat mobilizes cadmium plating	Select food contact surfaces carefully
Copper poisoning	Depends on dose (24–48 hrs)	Nausea, vomiting, diarrhea	Acid foods, foods contacting copper	Acid mobilizes copper	Select food contact surfaces carefully
Lead poisoning	Depends on dose	Metallic taste, abdominal pain, vomiting, diarrhea, black stools, oliguria, collapse, coma (also chronic effects)	Glazes, glasses, illicit whiskey	Lead dissolves in beverages and foods	Test glazes and glasses; avoid illicit whiskey
Mercury poisoning	Depends on dose	Metallic taste, thirst, abdominal pain, vomiting, bloody diarrhea, kidney failure	Treated seeds (fungicide); fish	Intentional; food chain	Eat only seeds intended for food
Zinc poisoning	Depends on dose (24–48 hrs)	Nausea, vomiting, diarrhea	Acid foods in galvanized containers	Acid mobilizes zinc plating	Select food contact surfaces carefully

Chapter 4

SOME BASIC CHEMICAL DIFFERENCES IN FOODS

What are little girls made of?
Sugar and spice and everything nice.
What are little boys made of?
Rats and snails and puppy dog tails.
What are little plants made of?
Water and air and sunlight fair.
What are little animals made of?
Plants and worms, anything that squirms.
What are real people made of?
Food and drink and brains that think.
That's what we are made of.
— Fred Deatherage, *Food for Life*, 1975

Foods are composed of water, carbohydrates, proteins, fats (lipids), minerals, vitamins, organic acids, hormones, pigments, flavones, essences, antioxidants, and many other chemical substances. These main chemical constituents are the major differences of most of our foods. These differences are the primary reason that we must vary our diet and utilize many types, styles and kinds of foods to obtain the necessary nutrients needed by the body.

Foods provide us with energy, foods promote growth and repairs damaged and worn out tissues. Food makes reproduction possible and the foods we eat provide unique esthetic and psychological effects. In times of stress and unhappiness, foods may be our best relief and provide much more than the above. Having spent some time in the hospital, I would tell you that my meals were something that I looked forward to with much anticipation. I knew that the many nutrients contained therein were needed to bring me back to good health and I knew that the meals were most important in my early and complete recovery.

Carbon dioxide, water, green plants and sunshine are the basic ingredients and the biological starting point for the manufacture of all of our food. Man and other heterotrophs utilize the substances from photosynthesis for survival. Biological and biochemical activities are all part of the eco-system as it is often called today. They are essential for man's survival.

Glucose is one of the first products of photosynthesis. Glucose is made up of carbon, hydrogen and oxygen. It is known as a simple carbohydrate and it is the building block for all other carbohydrates. The name carbohydrate implies that hydrogen and oxygen are present in the same ratio as they are in water, that is, 2:1. Carbohydrates are a major part of the composition of nearly all plant materials (75 percent of the dry weight of plants). Carbohydrates include the simple sugars, such as glucose, other sugars, starch, fiber, and cellulose. Carbohydrates provide energy (4 calories per gram) and they are the raw material for the manufacture of many of the B complex vitamins. Carbohydrates, also, add flavor to foods and provide some attractive colors in our cooked foods.

Animal products are made up of proteins and fats (lipids). In addition animal products contain large amounts of minerals, primarily in their skeletal tissues which we do not eat as such. However, both edible plants and animal tissues contain about 1% of mineral matter and many of these minerals are most essential. Deatherage published data on the proximate composition of some foods as shown in Table 4.1.

TABLE 4.1 — Proximate Analysis in Percent
for Typical Foods of Plant or Animal Origin*

Food	Protein	Fat	Carbohydrate		Ash	Water
			Total	Fiber		
Sweet Corn	3.5	1.0	22.1	0.7	0.7	72.0
Apple	0.3	0.5	15.0	0.9	0.3	84.0
Whole Wheat	13.3	2.0	71.0	2.3	1.7	12.0
Whole Egg	12.8	11.5	0.7	0.0	1.0	74.0
Lean Sirloin Steak	21.5	5.7	0.0	0.0	1.0	71.8
Cooked Hamburger	24.2	20.5	0.0	0.0	1.1	54.2

*Taken from Deatherage, *What Is Food Made Of?*, p. 113.

Lipids or fats and oils are what the chemist call esters of glycerol and fatty acids (one molecule of glycerol to three molecules of fatty acids). Glycerol comes from glucose and the fatty acids are built up from acetic acid (which also comes from glucose). The fatty acids vary in carbon chain length with the shorter chain fats coming from coconut and milks of mammals and the longer fatty acid chains coming from fish and peanuts. The most common fatty acids contain 16 to 18 carbon atoms in their chains. Fats vary widely in their chemical structure and they are solid at room temperature. Oils, also, vary widely in composition, but they are liquids at room temperature. All lipids are insoluble in water.

Fats that contain the maximum number of hydrogen atoms are known as saturated fats. Unsaturated fats do not contain all the hydrogen atoms possible (oils are goods examples). However, it is possible to add hydrogen to an unsaturated fat and make it a solid saturated fat. Fats are a concentrated source of energy providing 9 calories per gram. Vitamins A, D, E, K are carried in fat as they are fat soluble. (See Table 4.2 for the composition of some fats and oils.)

TABLE 4.2 — Sources of Fatty Acids in Some Unhydrogenated Fats and Oils

Fatty Acid	Soybean	Cotton	Corn	Peanut	Canola	Sunflower	Coconut	Palm
Caprylic 8:0						8.0		
Capric 10:0						6.0		
Lauric 12:0						48.0		
Myristic 14:0	0.1	1.0	0.1	0.1	0.1	18.2	1.0	
Palmitic 16:0	11.0	23.0	11.0	11.5	4.3	7.0	9.0	46.0
Stearic 18:0	3.6	3.0	2.0	2.0	1.7	4.2	2.0	4.0
Oleic 18:1	24.7	18.0	27.0	48.0	59.1	19.5	7.0	37.0
Linoleic 18:2	53.5	54.0	58.5	31.0	27.8	68.9	2.0	10.0
Linoleic 18:3	6.4	0.4	0.5	1.0	8.2			1.0
Arachidic 20:0	0.3	0.2	0.3	1.5	0.5	0.3		1.0
Gadoleic 20:1					2.0			
Behenic 22:0				3.0				
Euric 22:1					0.9			
Lignoceric 24:0				1.0				

Essential oils are found in the lipid fraction of many foods and they have a profound effect on the odors and flavors in many foods. Essential oils are, also, termed oleoresins or "essences". All together these oleoresins and essences may make up the flavor, odor, and acceptability of many foods. Again from Deatherage, the information in Table 4.3 through 4.7 shows the components in Tomato, Black Pepper, Onion, Orange, and Beef to illustrate the naturally occurring compounds in these foods. As one can see, there are many chemical components of any given product.

TABLE 4.3 — Some of the Compounds Naturally
Present in Tomato (Lycopersicum esculentum)

Decane	Penta-2-enal	y-Hexalactone
Undecane	Penta-3-enal	y-Octalactone
Benzene	Hex-*trans*-2-enal	y-Nonalactone
Toluene	Hex-*cis*-3-enal	2,2,6-Trimethyl-2-
p-Xylene	Hex-*trans*-3-enal	hydroxycyclohexylidene
Isopropyl benzene	Hept-*trans*-2-enal	acetic acid lactone
pseudocumene	Oct-*trans*-2-enal	Ethyl acetate
Limonene	Non-*trans*-2-enal	Butyl acetate
a-Pinene	Hexa-*trans-trans*-2,4-	Pentyl acetate
Myrcene	dienal	Hex-3-enyl acetate
Δ³-Carene	Hepta-*trans-cis*-2,4-	Methyl hexanoate
Methanol	dienal	Butyl nitrile
Ethanol	Hepta-*trans-trans*-2,4-	Isobutyl nitrile
Propanol	dienal	Phenyl acetonitrile
Isopropanol	Deca-*trans-trans*-2,4-	Hydrogen sulfide
Butanol	dienal	Dimethyl sulfide
Isobutanol	Deca-*trans-cis*-2,4-dienal	Dimethyl disulfide
Pentanol	Benzaldehyde	2-Methyl mercaptoethanol
2-Methyl butanol	Cinnamaldehyde	3-Methyl mercapto-
3-Methyl butanol	Hydroinnamaldehyde	acetaldehyde
Pent-1-en-3-ol	Phenylacetaldehyde	3-Methyl mercapto-
Hex-2-enol	Salicylaldehyde	propanol
Hex-*cis*-3-enol	Citral	2-Ethyl furan
Benzyl alcohol	Neral	2-Pentyl furan
2-Phenyl ethanol	Geranial	2-Acetyl furan
Phenol	Acetone	Furfural
Methyl salicylate	Butanone	5-Methyl furfural
o-Cresol	Pentan-2-one	2-2,4-Trimethyl-1,3-
Guaiacol	Pentan-3-one	dioxalene
p-Ethyl phenol	Pent-1-en-3-one	2-Isobutyl thiazole
p-Vinyl guaiacol	Hexan-2-one	Acetophenone
Eugenol	Non-*trans*-2-en-3-one	-Hydroxyacetophe-
Linalool	Diacetyl	none
Linalool oxide	Butan-2-ol-3-one	2-Methylhept-2-en-6-
2-Methylhept-2-en-6-ol	Penta-2,3-dione	one
a-Terpineol	Geranylacetone	2-Methylhepta-2-*trans*-
Acetaldehyde	Farnesylacetone	4-dien-6-one
Propanal	2-Methyl butanal	β -Ionone
Farnesal	3-Methyl butanal	Pseudoionone
Glyoxal	Acetic acid	Epoxy-5,6-ionone
Methyl glyoxal	Propionic acid	2,2,6-Trimethyl-2-
Hexanal	2-Methyl butyric acid	hydroxyhexanone
Heptanal	y-Butyrolactone	Pentanoic acid

TABLE 4.4 — Some of the Compounds Naturally
Present in Black Pepper (*Piper nigrum*)

a-Pinene	β-Farnesene	Terpinolene
β-Pinene	a-Humulene	Ocimene
a-Phellandrene	β-Bisabolene	Arcurcumene
β-Phellandrene	y-Muurolene	Epoxydihydrocaryo-
DL-Limonene	a-Selinene	phyllene
β-Caryophyllene	β-Selinene	Phenylacetic acid
β-Elemene	δ-Cadinene	Dihydrocarveol
δ-Elemene	Calamanene	Piperonal
a-Cubebene	a-Thujene	Cryptone
a-Copaene	Camphene	cis-p-Menthen-1-ol
a-cis-Bergamotene	Sabinene	cis-p-2,8-Menthen-1-ol
a-trans-Bergamotene	Δ³-Carene	trans-Pinocarveol
a-Santalene	a-Terpinene	Piperidine
Hydrocaryophyllene	Myricene	
Isocaryophyllene	p-Cymene	

TABLE 4.5 — Some of the Compounds Naturally
Present in Onion (*Allium cepa L.*)

Propene	Methyl cis-propenyl disulfide	Isopopyl propyl trisulfide
Propanal	Methyl trans-propenyl disulfide	Dipropyl trisulfide
Dimethylfuran	Isopropyl propyl disulfide	cis-Peopenyl propyl trisulfide
2-Methylpentanal	Dipropyl trisulfide	trans-Propenyl propyl trisulfide
2-Methyl-pent-2-enal	cis-Propenyl propyl disulfide	Dimethyl tetrasulfide
Tridecan-2-one	trans-Propenyl propyl disulfide	2,5-Dimethylth-iophene
5-Methyl-2-*n*-hexyl-2,3-dihydrofuran-3-one	Diallyl disulfide	2,4-Dimethylth-iophene
Hydrogen sulfide	Allyl propenyl disulfide	3,4-Dimethylth-iophene
Methanethiol	Dipropenyl disulfide	3,4-Dimethyl-2,5-di-hydrothiophene-2-one
Propanethiol	Dimethyl trisulfide	Methyl methane thiosulfonate
Allylthiol	Methyl propyl trisulfide	Propyl methane thio-sulfonate
Dimethyl sulfide	Allyl methyl trisulfide	Propyl propane thio-sulfonate
Allyl methyl sulfide	Methyl cis-propenyl trisulfide	
Methyl propenyl sulfide	Methyl trans-propenyl trisulfide	
Allyl propyl sulfide	Diisopropyl trisulfide	
Propenyl propyl sulfide		
Dipropenyl sulfide		
Dimethyl disulfide		
Methyl propyl disulfide		
Allyl methyl disulfide		

TABLE 4.6 — Some of the Compounds Naturally
Present in Roasted, Fried and Boiled Beef

n-Nonane	3-Methylbutanal	2-Furfurylmethylke-
n-Decane	n-Pentanal	tone
n-Undecane	n-Hexanal	Acetylpyrrole
n-Dodecane	2-Hexanal	2-Methyl-acetyl
n-Tridecane	n-Heptanal	pyrrole
n-Tetradecane	2-Heptanal	o-Hydroxyacetophe-
6-Methyltetradecane	n-Octanal	none
n-Pentadecane	2-Octanal	4-Hydroxy-5-methyl-
n-Hexadecane	2,4-Octadienal	3(2H)-furanone
n-Heptadecane	n-Nonanal	4-Hydroxy-3,5-
n-Octadecane	2 -Nonanal	dimethyl-3(2H)-
Toluene	2-4-Nonadienal	furanone
1,4-Dimethylbenzene	n-Decanal	Acetic acid
1,2-Dimethylbenzene	2-Decanal	Propionic acid
Trimethylbenzene	2,4-Decadienal	Butyric acid
Methylethylbenzene	n-Undecanal	Isobutyric acid
Diethylbenzene	2-Undecanal	Valeric acid
n-Butylbenzene	2-4 Undecadienal	Benzoic acid
2-n-Pentylfurane	2,4-Dodecanal	Lactic acid
2-n-Hexylfurane	n-Dodecanal	Acetol acetate
2-n-Octylfurane	2-Dodecanal	y-Butyrolactone
Dimethylpyrazine	2,4-Dodecadienal	Methylbutyrolactone
Trimethylpyrazine	n-Tridecanal	y-Hexalactone
Dimethylethylpyrazine	2-Tridecenal	y-Heptalactone
1-Propanol	n-Tetradecanal	δ -Heptalactone
1-Pentanol	n-Pentadecanal	y-Octalactone
1-Hexanol	n-Heptadecanal	y-Nonalactone
n-Butoxyethanol	Benzaldehyde	δ -Nonalactone
1-Heptanol	Phenylacetaldehyde	y-Decalactone
2-Heptanol	Methylcinnamaldehyde	2,4,5-Trimethyl-Δ³-
3-Heptanol	Butan-2-ol-3-one	oxaline
4-Heptanol	2,3-Butadione	Thiophene-2-carboxal
1-Octanol	2,3-Pentadione	dehyde
3-Octanol	2-Heptanone	5-Methiofurfuralde-
4-Octanol	3-Heptanone	hyde
2-Octen-1-ol	2-Octanone	Benzothiazole
1-Nonanol	3-Octanone	Dimethyldisulfide
1-Decanol	2-Nonanone	1-Methylthioet-
1-Undecanol	2-Decanone	hanethiol
1-Dodecanol	2-Undecanone	Methional
Phenol	2-Dodecanone	Dimethylsulfone
Benzyl alcohol	2-Tridecanone	2,5-Dimethyl-1,3,4-
Vinylguaiacol	2-Pentadecanone	trithiolane
2-Methylbutanal	Acetyl furan	

TABLE 4.7 — Some of the Compounds Naturally
Present in Orange (*Citrus sinensis,* Valencia)

Hexane	Ethanol	Octanal
Isoprene	Linalool	Nonanal
Methyl cyclopentane	Octanol	Decanal
Heptane	Nonanol	Neral
Octane	*trans*-2,8-*p*-Menthadien-1-ol	Geranial
Nonane	*cis*-2,8-*p*-Menthadien-1-ol	Dodecanal
a-Pinene	*a*-Terpineol	Perillaldehyde
Sabinene	Citronellol	Acetone
Myrcene	*trans*-Cerveol	Ethyl vinyl ketone
D-Limonene	*cis*-Carveol	Carvone
β -Cubebene	1,8-*p*-Menthadien-9-ol	Piperitenone
β -Elemene	8-*p*-Menthene-1,2-diol	Diethyl acetal
β -Copaene	Hexanal	Ethyl butyrate
Valencene	Heptanal	1,8-*p*-Menthadien-9-yl
		acetate

Data compiled from C.H.T. Tonsbeek, A.J. Plancken and T.V.D. Weerdhof, *J. Agr. Food Chem.* 16:1017 (1968); H.M. Liebich, D.R. Douglas, Albert Zlatkis, Francoise Muggler-Chavan and A. Donzel, *J. Agr. Food Chem.* 20:96 (1972); Kenji Watanabe and Yasushi Sato, *J. Agr. Food Chem.* 20:174 (1972); H.W. Brinkman, Harald Copier, J.J.M. Deleuw and sing Boen Tjan, *J. Agr. Food Chem.* 20:177 (1972); Stephen S. Chang, *Proceedings 26th Annual Conference, Am. Meat Sci. Assoc.* 76 (1973); Ira Katz, *Proceedings 26th Annual Conference, Am. Meat Sci. Assoc.* 102 (1973).

Protein comes from the Greek word meaning "holding first place" or "primary," and no biological system can live and no process can take place without a protein being involved. Proteins are fundamental to all aspects of cell structure and function. All cells contain proteins. Proteins are essential in nutrition and they provide 4 calories per gram.

Enzymes are protein in nature. Actually, enzymes do the controlling of all living processes. All proteins are made up of amino acids (protein building blocks) and they contain carbon, hydrogen, oxygen and nitrogen as primary elements. There are many amino acids, but there are 10 that are "essential" to man. They are: Arginine, Histidine and the following that are not synthesized in the body: Valine, Isoleucine, Leucine, Threonine, Methionine, Phenyllanine, Tryptophan, and Lysine. Proteins have pronounced effects on texture, tenderness, color and acceptability of various foods.

pH and the inherent acidity of our foods are most important attributes to classify foods, at least, from a processing standpoint as the pH tells us something about the protective properties of foods in terms of microbial growth and survival. pH is a scale used to evaluate the hydrogen ion concentration. It designates the degree of acidity or alkalinity of foods. The pH scale ranges from 0 (very acidic) to 14 (very alkalinic) (see Table 4.8—Relationship of pH Values to Concentration). In scientific language, pH is the negative logarithm of the hydrogen ion concentration. The pH is directly related to the ratio of hydrogen (H+) to hydroxyl (OH–) ions present in the food. pH of 7 is considered neutral to the chemist, but to the food processor and/or food technologist a pH of 4.6 is considered neutral in terms of effects on processing.

TABLE 4.8 — Relationship of pH Value to Concentration
of Acid (H+) or Alkalinity (OH–)

phH Value	Concentration	
0	10,000,000	
1	1,000,000	
2	100,000	
3	10,000	Acidity
4	1,000	
5	100	
6	10	
7	0	Neutral
8	10	
9	100	
10	1,000	
11	10,000	Alkalinity
12	100,000	
13	1,000,000	
14	10,000,000	

That is, foods with a pH of 4.0 do not normally require a pressure cooking process because the natural inherent acidity will not support the growth of public health-type microorganisms (see Table 4.9).

A pH meter should be a standard requirement for all food labs and it is a most essential tool for any one developing or formulating foods. Understanding and taking advantage of pH information can be most useful in food plant operations and food processing know-how (see Table 4.10).

TABLE 4.9 — Fruits and Vegetables
Classed According to Acidity

Group No.	Group Description	pH	Examples of Food Products
I	Non-acid	7.0–5.3	Corn, lima beans, peas
II	Low or medium acid	5.3–4.6	Beets, pumpkin
III	Acid I	4.6–3.7	Apricots, pears, tomatoes
IV	Acid II	3.7 & below	Apple sauce, grape-fruit, pickles

Another important chemical constituent of all foods is the water content and water activity level. This information is most valuable in terms of food qualities, food composition, food processing activities, and/or food shelf life. The many types of foods vary widely in water content and are listed in Table 4.10. Even within a given type or commodity, such as, potatoes the water content may vary *nearly* 20 percent due to cultivars, area of production, maturity, and/or storage conditions. The modern food processor uses water content as a measure of quality. In drying foods, water measurement is mandatory to assure quality and a safe process.

Since 1940, food technologists have understood that it is not the amount of water in a food that controls microbial growth, but rather the available water within the food material that greatly influences microbial growth. The measurement of the available water or the A sub w value can be most helpful in predicting microbial growth. Molds will grow with an A_W value of 0.75, yeast at 0.88, and *Clostridium botulinum* at 0.93. Water activity levels can be determined quite easily using a sensor to measure the equilibrium relative humidity.

There are many other constituents of food that have a direct or indirect bearing on food processing, food quality, and food acceptability. However, the above should inspire in one the necessity to understand basic biochemistry when working with foods and in particular when processing foods to assure uniformity in product quality. The biochemistry of foods is an exciting part of understanding the fundamentals of food processing and technology.

TABLE 4.10 — pH Values of Some
Commercially Canned Foods

	pH Values		
Canned Product	Average	Minimum	Maximum
Apples	3.4	3.2	3.7
Apple Cider	3.3	3.3	3.5
Apple Sauce	3.6	3.2	4.2
Apricots	3.7	3.6	3.9
Apricots, strained	4.1	3.8	4.3
Asparagus, green	5.5	5.4	5.6
Asparagus, white	5.5	5.4	5.7
Asparagus, pureed	5.2	5.0	5.3
Beans, Baked	5.9	5.6	5.9
Beans, Green	5.4	5.2	5.7
Beans, Green, pureed	5.1	5.0	5.2
Beans, Lima	6.2	6.0	6.3
Beans, Lima, pureed	5.8	–	–
Beans, and Pork	5.6	5.0	6.0
Beans, Red Kidney	5.9	5.7	6.1
Beans, Wax	5.3	5.2	5.5
Beans, Wax, pureed	5.0	4.9	5.1
Beets	5.4	5.0	5.8
Beets, pureed	5.3	5.0	5.5
Blackberries	3.6	3.2	4.1
Blueberries	3.4	3.3	3.5
Carrots	5.2	5.0	5.4
Carrots, pureed	5.1	4.9	5.2
Cherries, black	4.0	3.9	4.1
Cherries, red sour	3.3	3.3	3.5
Cherries, Royal Ann	3.9	3.8	3.9
Cherry Juice	3.4	3.4	3.4
Corn, W.K., brine packed	6.3	6.1	6.8
Corn, cream-style	6.1	5.9	6.3
Corn, on-the-cob	6.1	6.1	6.1
Cranberry Juice	2.6	2.6	2.7
Cranberry Sauce	2.6	2.4	2.8
Figs	5.0	5.0	5.0
Gooseberries	2.9	2.8	3.2
Grapes, purple	3.1	3.1	3.1
Grape Juice	3.2	2.9	3.7
Grapefruit	3.2	3.0	3.4

Continued on next page.

TABLE 4.10 — pH Values of Some
Commercially Canned Foods - (Continued)

| Canned Product | pH Values | | |
	Average	Minimum	Maximum
Grapefruit Juice	3.3	3.0	3.4
Lemon Juice	2.4	2.3	2.6
Loganberries	2.9	2.7	3.3
Mushrooms	5.8	5.8	5.9
Olives, Green	3.4	–	–
Olives, ripe	6.9	5.9	7.3
Orange Juice	3.7	3.5	4.0
Peaches	3.8	3.6	4.1
Pears, Bartlett	4.1	3.6	4.7
Peas, pureed	5.9	5.8	6.0
Peas, Alaska, (Wisc)	6.2	6.0	6.3
Peas, sweet wrinkled	6.2	5.9	6.5
Peas, pureed	5.9	5.8	6.0
Pickles, Dill	3.1	2.6	3.8
Pickles, fresh cucumber	4.4	4.4	4.4
Pickles, sour	3.1	3.1	3.1
Pickles, sweet	2.7	2.5	3.0
Pineapple, crushed	3.4	3.2	3.5
Pineapple, sliced	3.5	3.5	3.6
Pineapple, tidbits	3.5	3.4	3.7
Pineapple Juice	3.5	3.4	3.5
Plums, Green Gage	3.8	3.6	4.0
Plums, Victoria	3.0	2.8	3.1
Potatoes, Sweet	5.2	5.1	5.4
Potatoes, White	5.5	5.4	5.6
Prunes, fr. prune plums	3.7	2.5	4.2
Pumpkin	5.1	4.8	5.2
Raspberries, black	3.7	3.2	4.1
Raspberries, red	3.1	2.8	3.5
Sauerkraut	3.5	3.4	3.7
Spaghetti in Tomato Sauce	5.1	4.7	5.5
Spinach	5.4	5.1	5.9
Spinach, pureed	5.4	5.2	5.5
Strawberries	3.4	3.0	3.9
Tomatoes	4.4	4.0	4.6
Tomatoes, pureed	4.2	4.0	4.3
Tomato Juice	4.2	4.0	4.3

TABLE 4.11 — Moisture Content in Fresh
Fruits & Vegetables

Product	Average	Maximum	Minimum
Apples	84.1	90.9	78.7
Apricots	85.4	91.5	81.9
Avocados	65.4	68.4	60.9
Blackberries	85.3	89.4	78.4
Cherries, sweet	80.0	83.9	74.7
Figs	78.0	88.0	50.0
Grapefruit	88.0	93.1	86.0
Grapes, European	81.6	87.1	74.8
Muskmelon	92.8	96.5	87.5
Oranges	87.2	89.9	83.0
Peaches	86.9	90.0	81.9
Pears	82.7	86.1	75.9
Prunes, fresh	76.5	89.3	61.6
Rhubarb	94.9	96.8	92.6
Watermelons	92.1	92.9	91.3
Artichokes	83.7	85.8	81.6
Asparagus	93.0	94.4	90.8
Beans, Lima	66.5	71.8	58.9
Beans, Snap	88.9	94.0	78.8
Beets	87.6	94.1	82.3
Cabbage	92.4	94.8	88.4
Carrots	88.2	91.1	83.1
Cauliflower	91.7	93.8	87.6
Celery, stalks	93.7	95.2	89.9
Corn, sweet	73.9	86.1	61.3
Cucumbers	96.1	97.3	94.7
Lettuce	94.8	97.4	91.5
Onions	89.2	92.6	80.3
Peas, green	74.3	84.1	56.7
Potatoes	77.8	85.2	66.0
Pumpkin	90.5	94.6	84.4
Spinach	92.7	95.0	89.4
Tomatoes	94.1	96.7	90.6

SOURCE: Part from Joslyn (1950).

Chapter 5

FOOD PLANT LOCATION

"The best thing about the future is that it comes one day at a time"
— Dean Acheson

Food manufacturing facilities are located in every state in the U.S. and in many foreign countries. Food processors in the U.S. have been the leaders in food preservation.

Food factories used to be located in the rural areas, but today they may be found within the city limits as many food plants are termed secondary food processors. That is, they re-manufacture the original processed products into complete meals, specialty items, drinks, and many entree items. However, all of our food originally comes from the cultivated land or water and, therefore, rural areas are the usual starting point of any food operation.

Location of a food factory is most critical for those commodities that can deteriorate on holding prior to processing. One commercial ad states that their products are processed within 30 minutes from harvest to packaging. The days of the long cattle drives or wagon loads of produce lined up on the highway waiting to be processed are gone. Many processing factories are located in the area of production to handle the products promptly from harvest and to preserve what nature made.

In locating a food factory there are many factors that should be investigated before purchasing the land and/or building the factory. The following are my suggestions:
- Community or municipality acceptance
- Available raw materials including ingredients, packages, etc.
- WEWEL, that is, Water supply, Energy, Waste disposal, Environment, and supply of Labor
- Transportation
- Marketing area
- Available Technology
- Business Climate including available Money

The list could be added to, but the above are my primary concerns.

COMMUNITY ACCEPTANCE

Community or the potential municipality is first on my list because one needs to be satisfied that the community wants a factory in their area. Having served as a City Councilman, I know of some of the outcrys and put-ons. The "Not In My Back Yard" (NIMBY) syndrome is always present and it takes much public relations to satisfy that alarm. One only needs to see how some city councils act when they know the problem. I have seen major firms having to close their factories because of the aroma from a tomato manufacturing plant, the steam from a chip plant when frying potatoes, the odor of an animal slaughtering plant, etc. It behooves a food firm to make available any potential environmental problem at the start of negotiations rather than build a factory and have to close because of environmental factors. The environment and any potential pollutants or contaminants should be the very first concern of every food processing operation. Every food factory or for that matter every business wants to feel welcome.

My first suggestion is that the President or CEO take the time to visit with the city leadership and inform them of the proposed plans, any forseable potential problems and the opportunities that could lie ahead. I served one time as the business recruiting councilman for my city and I did not wait for a firm to come to the council or city government. I visited several firms interested in moving their office and plant site, even in other states to understand their operation. A firm should feel welcome because of the taxes it may pay, the employment opportunities, the dollars for other businesses in town, etc. On the latter, I know of a Governor that publicized the fact that for every dollar the food firm generated, there were 7 other dollars left in the community—gasoline, tires, meals, clothing, homes, taxes, etc. A business is a great asset to many communities and it may aid in keeping the community more tax free. Everyone must work together and understand the pros and cons of a new business in town and work for the acceptance of the potential firm.

RAW MATERIALS

A food plant is generally in business to preserve food that may be very perishable or to add value to other raw materials or secondary products. Food plants are generally located at the source of the raw materials. One does not usually build an orange juice factory in Ohio when oranges are indigenous to Florida, Texas or California. However, one can build a factory in Ohio to add value to oranges and orange products that have been grown and first processed elsewhere. The Ohio plant, for example, may only be repackaging the orange juice from tanker deliveries or the plant may be blending the orange juice with other juices, packaging it and selling direct or indirect to the customer.

The primary factory may be just a bulk manufacturing facility with full expectations of selling their product for re-manufacture. This is happening with tomatoes and some other commodities these days wherein the paste or other products made at the source of production are sold to plants far distant from the producing area. For example, the receiving plant re-manufactures the paste into tomato and vegetable juices, dressings, sauces, salsas, soups, and ketchups. These re-manufactured products add much value to the new commodities that are finding wide usage with the customer.

Some products like grains, potatoes and other root crops can be stored and/or shipped long distances as needed (Just in Time (JIT)) and be very highly acceptable to the end use manufacturer. Potatoes, for example, are shipped hundreds of miles for chip usage, likewise grains used in the manufacture of cereal, breads, and snacks.

Packaging materials and other ingredients are essential raw materials for any processing or food preservation facility. They must be available at reasonable prices and they must be available as needed. The package may be as expensive as the raw material, depending on the size of the package and the style and type. Nevertheless, packaging is a major component of the food business.

Whether we like it or not, today the food industry is basically a packaging industry and packages must be reckoned with when locating a food facility. No one really wants to go back to the good old days of the cracker barrel. Modern, attractively packaged foods are part of this industry and packaging is here to stay. Packages and packaging should protect the food product from human tampering, insects and rodents, air, moisture, and/or light.

WEWEL or Water, Energy, Waste, Environment and Labor

This category on my list is so fundamental to any food business that it is probably not necessary to mention these items. However, the following may bring to mind the need and some considerations for each of these items.

Water

All segments of the food industry use water to clean the raw food material and to clean the factories and, yes, to clean personnel working in the factory. Water is used to clean the raw materials, used in blanching food, used in cooling of produce and finished canned products, used in conveying produce into and through the factory, and used for the generation of steam used in processing. In these times it is most gratifying to learn that the meat industry is using water to wash the carcass prior to cutting etc. The canning industry uses water in the brine and syrups to help, in part, in the energy transfer to provide adequate sterilization of the food.

Water is important and the quality of that water must be ascertained when locating a facility. Water for food processing should be potable, soft, and free of any microbes and minerals. Sure, water can be filtered, chlorinated, and de mineralized, but why do these things when it may be fully acceptable as is. One must know, however, to be sure.

The amount of water varies with the commodity, but a general rule of thumb is that it takes up to 2 1/2 gallons of water to prepare and process each pound of raw material. Of course, this varies depending on whether you are a primary processor or a secondary processor and with the commodity you are processing.

Energy

In the old days, we burned wood, coal and oil for our sources of energy and all was satisfactory until we over polluted the environment. Today, we generally use natural gas and we have eliminated much of the pollution and provided a satisfactory source of energy. However, we have a long way to go to satisfy some people in providing other alternatives and/or relative inexpensive sources of energy. In my opinion the time will come when we will fully rely on solar and/or nuclear energy or similar sources to supply our energy needs.

Any city or municipality that can provide and sell these sources of energy at an economical price will have a great incentive to attract food processing facilities. Energy is the heart and soul of most food processing operations.

The amount of energy varies widely because of the uses by the canner, the freezer, the dehydrator, and or the snack food manufacturer. The only food preservation industry that is not a major user of energy is the fermentation industry. They utilize the natural energy produced by the microorganisms and the energy stored in the fermenting product.

There are indications on the horizon for using more electric energy including ohmic heating to sterilize our foods. Time will only tell how well we improve the existing means of energy sources and energy deployment to solve a much needed energy problem for every food processor. In a food processing plant, energy must be available, it must be relatively inexpensive, and it must be clean and not the cause of any pollution.

Waste

Frankly, there is little need for the amount of waste produced by the food industry. Much of that waste should and could be utilized as food. Tradition and regulations are part of the problem. Just think if we didn't peel potatoes, tomatoes, beets, carrots, etc.? What if we didn't remove the blossom end on green or wax beans, or the pods on lima beans and peas? What if we didn't peel the apple, the orange, the grapefruit, the pear, and the banana? What if the plant breeder or genetic engineer developed new products that had no waste to remove or throw away? To me, a great opportunity lies ahead in learning to cope with the utilization of our present waste through new manufacturing practices, such as, fermentation and/or the development of new products. The late J. R. Geisman developed some excellent edible food products from tomato peel, tomato seeds and tomato fiber. Tomato seeds are one of the best sources of complete protein available. An apple processor in this area of USA has found a tremendous market for dried pomace from his juice operation. Other examples could be cited, but this should point up a major need for all of us to look to the future in handling our wastes.

Factory wastes are on average 10 times stronger than domestic wastes. Some of this waste comes from the raw material and they may be sent back to the grower and he can add them back to the land or bury and cover these wastes. Liquid wastes are a major part of wastes and they include clean up waters, cooling waters, flumes and washer waters, and the blancher and peeler waters.

Wastes are disposed of in many ways, including burying or land fills, burned for energy at the factory site, digestion in ponds or compounding areas, fed to livestock or fish, or spread back on the land or composting. There is no easy system, but it must be a factor to consider when locating a factory.

The food plant operator must know the suspended solids in the water and the Biological Oxygen Demand (BOD) in addition to the pounds of waste and/or the gallons of waste waters. Waste from a food operation is an integral part of doing business today, however, utilization of this waste is a real challenge for the future.

Environment

The food industry certainly has contributed its share in creating problems with the environment, but certainly the problems can and should be turned into opportunities. No one will argue that an unsightly environment or a polluted environment is desired. Everyone will agree that the environment should be clean, odorless, and noise free.

Let me give you one example that I think should not be a problem. I have a neighbor that has a condo on the canal in back of my house. He has a large diesel boat that at idle speed rattles the dishes on our table when he goes out or comes back in to say nothing of the noise. Complaints have been filed, but until there are regulations established, the city fathers say nothing can be done. This is pollution to me and it should not be permitted.

Many pollution problems are exactly like this, to the polluter they are not problems; but, to those affected they are real pollution problems. In this case noise and vibration are pollution problems. Sure I could wear ear plugs or better still I could move. We will be watching this in the years to come and see just what happens.

Labor

At one time many of us worked in factories to process the seasonal crops. Gradually, man decided that this work was too tedious, too hard, too inconvenient with too little pay. The result was the drafting of migrants and "off-shore" labor to do the work. Now we find that labor is reluctant to work without lifetime contracts for jobs for their off-spring. So, what is management to do. Some management people are resorting to automation, robotics, and highly technical computer operated equipment. The end result is loss of many former jobs, but with more hand-outs for those that are not trained or for those unable to find the right job.

The food industry will survive and it may be better off as it learns to update its operations and becomes much more efficient and productive. Machines do not take breaks, they do not need vacations, and they do the same thing day in day out to assure quality. Labor will be the looser and you and I may pay more for our products, but we may be more satisfied because we get what we expect everytime when we buy that product. I'm not anti labor, but I am anti some of the things that labor does or tries to do and, hopefully, the food industry will not be one of the targets that some labor groups are after.

There are great jobs in the food industry and labor needs to be part of this industry. Labor needs to understand team building and team work and management must learn to keep labor informed and up to date. Management must spend more time and money to help labor stay abreast of the demands of the job. Management must provide the right tools for labor to do their job effectively. Management must constantly evaluate where we are and where we want to be and management must help labor to do their job effectively and to constantly improve productivity. Working together is the only way to continue to move forward.

Transportation

Today the food industry is heavily reliant on trucks for much of the transporting of raw and finished products. Some firms own their own fleet while others rent or contract for hauling. There are some items that still move by rail and air. Regardless of the method of transport, the hauler should understand the effects of time, temperature, air movement, relative humidity, and/or other gases or possible pollutants on the flavor, texture, and color of the food materials. Food firms rely on Just In Time (JIT) deliveries and shipments and the hauler must understand his role in the long chain of moving raw materials to the factory and from the factory to the customer. Transportation is a costly part of any food operation and everyone from the loader to the unloader must do their part, including the driver.

Cleanliness of the vehicles is one of the most important aspect of any movement of materials to and from the factory. The food industry must learn to police itself before more rigid regulations become the law of the land. With HACCP and the CGMP's already in place, one must caution this segment of a food firm to be alert, reliable in being safe in handling food materials and finished products and always stay up-to-date.

Technology

The above discussion of my views of WEWEL and the food industry leads me to the technology side of this great industry. To say the least the food industry has been slow to move into the technological era. The future awaits those that get up and get going. Equipment and procedures are available to help those that help themselves. For example, have you ever served as a food inspector in a factory looking for defects on products moving at the rate of 20' per minute. I have and I get dizzy, my eyes become blurred, and I know I did a poor job of removing the unwanted from the wanted. Its not a pleasant or an endearing job.

Today, we have equipment that will detect and kick out or remove the defective items without any one attending the operation. The machines and systems work and could be most beneficial in giving the customer what they expect all the time. However, I know of one firm that installed such equipment and did not like it because it removed more defects than management thought were in the products. Obviously, they didn't understand adjustments and they blamed the machine because of lack of know how on their part.

The food industry needs to become more technically inclined and learn to utilize new technology that can help management improve its operation, eliminate the defects, and become much more efficient. Personnel must stay up or move ahead of existing technology. Believe it or not, people want to learn, if given that opportunity. The future is most challenging for all concerned.

MARKETING AREA

There really should be no limit to the marketing area today as we should create conditions whereby the buyer/user convinces him or herself to want your product. Marketing is most critical and any firm in the food business must create a demand and/or convince the customer you have what they need and want. All of this should be developed before you even think of building or developing a food facility.

All firms should start small and build as they grow their market. There are too many food firms that want to be large overnight and act like the big boys with an already established reputation. One must earn that right and it takes planning, developing, synthesizing, and doing what is right all the time to grow a firm.

Success is and should be built on your repeat business because of the reputation you have established and upheld.

BUSINESS CLIMATE

One must realize up front that the business climate may not always be what you hope for, therefore, well laid plans may not come to fruition overnight. The business climate must be on your side when you start a new operation. You need capital, you need leadership, you need knowledge, and you need the desire to build a business. Most importantly, you need enthusiasm and you need the willingness to give of yourself to build that business. Yes, you need pioneering expertise and the ability to communicate with all who will listen. You can do it, but do you have the drive, the health, the time and the funds to pull it all together.

Sometimes a partner or a consultant may be the way to make it. One of the first things I would do if I were going to start a new business would be to take the time to know the industry, know some leaders, attend some conventions and shows, and become thoroughly familiar with the markets before I jumped. I certainly would make acquaintances with the Land Grant University in my area and find their expertise and by all means I would work with the food processing industry association in the state or area for their inputs and all.

Building that new food business can be done and it may be the most rewarding experiences you will ever know. Entrepreneuring is a wonderful experience and saying its MINE is a great recognition for your efforts. Remember, all big firms today were little firms yesterday. Those that lead and venture forth are the ones to reap the rewards. You can succeed if there is a will. Remember, your goal is your dream with a deadline.

Some Questions Before Starting To Build

1. Does the future business indicate that the capital cost can be recovered in a reasonable time?
2. Do we have adequate finances or is new money available?
3. Do we have sufficient time to build the new factory or could we find a co-packer without giving away our product?
4. What provision should we make for expansion?
5. Can we find a competent architect that knows about the requirements of a food factory such as cleanliness, lighting, floor finishes, power demands, safety and fire protection, maintenance, ventilation/air conditioning?
6. Do we have available a General Contractor that understands our plans and is interested in seeing the project to fruition?
7. If the new plant is an extension of our present operation, are we utilizing the present space adequately and are we allowing room for future growth?
8. Is new machinery readily available and will it allow for future growth?
9. Have we considered the effect of the new site on present personnel and their willingness to relocate or travel the new distance from their present home?
10. What is the prime time to start construction and what are the schedules for completion, installation of equipment, and start-up?

Chapter 6

FOOD PLANT EQUIPMENT AND MACHINERY

"The way a team plays as a whole determines its success. You may have the greatest bunch of individual stars in the World but if they don't play together, the club won't be worth a dime."
— Babe Ruth

During the last 50 years, food processing has moved from an art to a highly sophisticated scientific operation. The hand methods of past years have changed to highly mechanical, physical, or chemical practices. With these changes, our food is superior in quality, it is safer, it is much more economical, and it is much more flavorful and nutritious. The consumer is the ultimate winner and they get what they expect everytime under a given label.

Commercial food processing is essentially a sequence of unit operations (steps) designed to prepare, package, and process the various food items. Obviously, the unit operations will vary widely depending primarily on the type of commodity being processed.

For purposes of understanding food processing operations, a unit operation may be defined as a given physical step in the preparation, packaging, processing or manufacture that is incapable of division into smaller units. It is a single step in the preparation, handling, and preservation of a given food item. The equipment for each unit operation in a food plant may be as different as there are commodities to be preserved.

There are several factors that determine the kind and type of equipment to be used, including the following:

1. The particular fruit, vegetable, grain, fish and/or the type of animal product to be processed.
2. The volume contemplated.
3. The quality, type, and style of pack desired.
4. The finances of the owner.

All equipment in a food factory shall be so designed from a sanitary standpoint including ease of thorough cleaning after use and of such material and workmanship as to be adequately cleanable, and all equipment shall be properly maintained (CFR 110.40). Secondly, all equipment should be economical to operate. Thirdly, all equipment should allow for continuous operation. Fourthly, all equipment should be maintenance free. Lastly, all equipment should be capable of efficient operation and be non destructive or be non wasteful of the food.

When purchasing any equipment for a food plant, the procurement personnel should ascertain the details of operating the equipment when installed, that is, the operating parameters—speed or running time and any specific dwell time for given quantities, temperature limits and tolerances, reaction, if any, to any chemicals, maximum and minimum pressures and/ or vacuums, lubrication periods, suggested through-put, etc. He should, also, obtain the details of clean-up procedures, hopefully cleaning in place (CIP), and any other sanitation practices in terms of food contact surfaces, joints, fittings, and dead ends. Further, all equipment shall be in compliance with the CGMP's.

To start with, I am assuming that the given commodity has been grown, harvested and delivered to the food processor to meet the processors specifications for acceptance. These specifications should detail not only the quantity and time of delivery, but the quality and safety of the received product.

It behooves all processors to sample loads for acceptance of quality, but, also, for pesticide residue analysis. There should never be a problem if the producer has used pesticides according to regulations and label instructions for each given commodity including proper amount or concentration of sprays or dusts or fumigants or maturing agents and approved time or number of days of application prior to harvest.

In every food plant one will find a limited piece of equipment that controls the volume of material being put through the plant. This may be a washer, a snipper, a cutter, a fryer, a packaging machine, a cooker or what have you. Once a person learns to recognize this limiting unit operation, he can then ascertain the approximate capacity of the factory and quickly determine the efficiency in terms of actual through put. The productivity of the operation is based on how well the unit operations are producing as a whole or how well the factory is operating in terms of producing at capacity. Most importantly, productivity is all about continuous improvement.

For every type of food material being prepared, packaged and processed, one should develop a flow chart. These flow charts should clearly show the sequence of unit operations as the product moves through the factory.

A flow chart should serve, at least, three purposes:

(1) A picture of the process for the employee to "see the trees within the forest." This can serve as a training step for the employee, that is, a step to expound upon the function of this unit operation and the work requirements for this unit operation, that is, a way of designating the parameters of operation and requirements for the given procedures,

(2) A method of listing the quality control limits and/or adjustments that are permitted in the flow of the product for that unit operation, and

(3) A description of any hazards, if any, for that given unit operation or process step along with the controls and/or preventive measures, that is, the critical control limits and their controls.

Thus, the flow chart provides a general picture of the process, that is, its flow through the factory. More importantly, the flow chart provides information as to the hourly volume or through put, and it could indicate the number of operators and labor required. If one knows the number of days of operation and the peak volume, one can ascertain the raw materials needed, the storage requirements, and establish the inventory policy for the given item.

In establishing a flow chart, one must know the sequence for each unit operation as the chart is only a schematic representation of the plant. Each piece of equipment has its own specifications as to the following: floor space, services, and material hook-ups per unit operation from the preceding and the subsequent unit operation.

In Figure 1, I have shown a generic fruit and vegetable flow chart. On this chart I have listed the Critical Control Points and the kinds of hazards that might be present in a typical operation. I have not enumerated the in-plant hazards, such as cleaning chemicals, lubricants, people, etc. These will be disucssed in the Sanitation Chapter No. 10.

FIGURE 6.1 — Generic Fruit and Vegetables Processing
Flow Chart with Critical Control Points (CCL's)

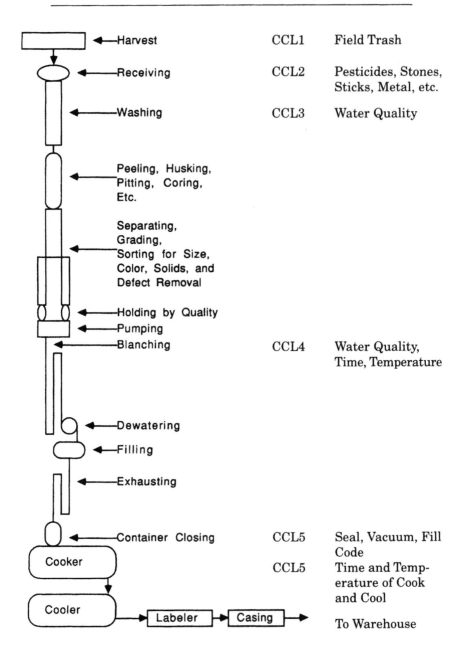

Harvest	CCL1	Field Trash
Receiving	CCL2	Pesticides, Stones, Sticks, Metal, etc.
Washing	CCL3	Water Quality
Peeling, Husking, Pitting, Coring, Etc.		
Separating, Grading, Sorting for Size, Color, Solids, and Defect Removal		
Holding by Quality		
Pumping		
Blanching	CCL4	Water Quality, Time, Temperature
Dewatering		
Filling		
Exhausting		
Container Closing	CCL5	Seal, Vacuum, Fill Code
Cooker	CCL5	Time and Temperature of Cook and Cool
Cooler		
Labeler	Casing	To Warehouse

Figure 2 is a specific flow chart for tomato juice manufacture. I have shown the quality control points in terms of suggested measurements. In addition, I have shown the Critical Control Points where applicable for each unit operation.

The flow chart should be helpful to management, it should be helpful to the new employee, and it should be most helpful to point up productivity problems. Ideally, the flow chart should be developed on the computer with the operating parameters, specific production data, and daily productivity calculated on a regular routine basis. Flow charts are helpful and most informative if properly utilized. Flow charts are the starting point in the design and layout of a factory.

Most importantly, a flow chart is a great training aid for use by new and experienced personnel to visualize the process and the specific use and function and how each unit operation contributes to the process. I have seen large drawings of flow charts displayed on walls of the training rooms to help orient and justify the unit operation. As new information comes forth or as changes are being studied, everyone can visualize the 5 W's and 1 H (What, When, Why, Where, Who and How) of the unit operation and/or the whole process. In other words, a good flow chart becomes a road map of the processing operation with all the stops indicated.

In my book, *Unit Operations for the Food Industries*, I have given specifics on many of the unit operations in a food plant, such as: Materials Handling, Cleaning, Quality Separation, Peeling, Disintegration with Little Change in Form, Disintegration with Considerable Change in Form, Separation, Blanching and Pre-cooking, Pumps and Pumping, Mixing and Blending, Additives and Coatings, Exhausting, Vacuumizing, Fillers, Closures, etc. Each of the many unit operations are defined and the objective(s) of the operation along with the methods in vogue today. Obviously, new changes constantly come to the fore front, but one needs to start with the basics and that is what one finds in this book.

Machines and equipment in a food plant must be constantly evaluated for efficiencies in terms of doing the expected job and their ability to be productive. Further, they must not allow any contamination of the product being processed. Management should constantly study the machines and the various operations to assure that one is doing what is expected all the time.

FIGURE 6.2 — Tomato Juice Processing
Flow Chart with QA‡ and CCL's*

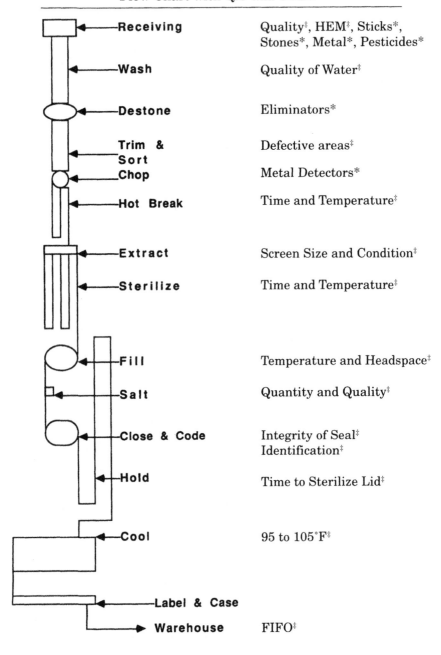

Process	QA / CCL
Receiving	Quality‡, HEM‡, Sticks*, Stones*, Metal*, Pesticides*
Wash	Quality of Water‡
Destone	Eliminators*
Trim & Sort	Defective areas‡
Chop	Metal Detectors*
Hot Break	Time and Temperature‡
Extract	Screen Size and Condition‡
Sterilize	Time and Temperature‡
Fill	Temperature and Headspace‡
Salt	Quantity and Quality‡
Close & Code	Integrity of Seal‡ Identification‡
Hold	Time to Sterilize Lid‡
Cool	95 to 105°F‡
Label & Case	
Warehouse	FIFO‡

Chapter 7

FOOD PACKAGING

"The path of Civilization is paved with tin cans"
— Elbert Hubbard

The package and the packaging of our food is a product of modern times. It may appear to be costly, but packaging of our food is most important from the standpoint of safety and prevention of people tampering. The modern day package may represent one-fifth to half the cost of the food depending upon the size of the package and the type of food. This is considered a good deal by most people because of what the package does do for the food.

The package should protect the product from contamination. The package should make the product acceptable and the package should present the food product to the customer attractively. Food packaging is a major part of the food industry and it is almost mandatory today for the safe supply of our food. The food preservation industry is basically a packaging industry.

Animal hides and horns, clay pots, and stoneware are some of the oldest containers used by man for his water and food. Glass was, also, one of the first commercially available containers. Glass is still used, although blowing the glass for mass production has changed to annealed containers. Blown glass containers has many advantages in that it can be personalized, that is, blown into various shapes and designs as desired by the buyer. Glass is desired for many products because of the man made colors by using different chemicals in the mix.

The can is without doubt the most popular container and it has stood the test of time for over 100 years. Plastic and flexible containers are the newest and offer some advantages that neither glass nor metal cans can.

Suggested Requirements for Food Packages:

1. All food containers should be non-toxic to the food, they should not react with the food, and they must not impart any flavor (taste or aroma) to the food.
2. All food containers should be moisture and vapor/odor proof and/or barrier resistant.
3. All food containers should be designed for complete closure and be easy to open and reclosure and be easy to remove the product from the container.
4. All food containers should be sanitary in design, easily cleanable before filling and they and their seal should protect the food from people tampering, and insect and rodent infestation.
5. All food containers should protect the food from light penetration and/or discoloration of the product.
6. Food containers should be resistant to impact and supportive of the product as to mechanical damage, that is, protection from crushing or breakage of the food product.
7. Food containers should be easy to dispose of and, hopefully, be biodegradable in a relative short time, or the food container should be recyclable.
8. The food container should be light in weight and unbreakable.
9. The food container should be available in different sizes.
10. The food container should be economical and relatively inexpensive.
11. The food container should be printable or designed for appropriate labeling and coding.
12. The food container should be convenient and easily handled by the customer including microwave heating whenever possible.

Types of Food Containers:

Food containers can be categorized as follows:
A. Metal Cans
B. Glass Bottles, Jars, and Jugs
C. Plastic
  1. Rigid
  2. Semi Rigid and/or Flexible
  3. Films

D. Paper, Cellulose and wood Containers
E. Edible Coatings

Metal cans

Metal cans became available back in the early 1800's as indicated in Chapter 2. They have changed over the years with the modern can fully opened at the top to allow proper cleaning before filling and for ease in filling. Further, the tin coating has become somewhat less and there is a total reduction of lead used to solder the cans and lids. The can today may be electrically welded or it may be a two piece container that is drawn and redrawn (DRD) or the drawn and ironed from a glob of steel. Generally, the inside of the container is coated with tin or with an organic compound to prevent chemical reactions between the food and metal.

The secret of the modern tin can is a rubber compound used between the metal in the cap and the metal of the body of the can to allow for a hermetic seal (defined as a container which is designed and intended to be secure against the entry of microorganisms and to maintain the commercial sterility of its contents during and after processing). The lid is still crimped or rolled in place as shown in Figure 7.1. The integrity of the seal may be the weakest link in the use of the metal container. However, it should never be a problem if the food processor is operating the closing machine properly.

The Code of Federal Regulations (CFR) requires that container closure inspections shall be made immediately following a severe jam at the closing machine and inspections shall, also, be made after a closing machine has been changed from one can diameter to another. The metal can is a most acceptable container for canned foods as it will take the strain of high temperatures during the processing of the food without breakage if properly filled and exhausted prior to closing. The metal can will withstand a certain amount of rough handling, but sharp blows to the seamed end may cause serious problems. Under normal handling practices this container is the ideal container for handling of most foods. It is the most important package and it continues to withstand the test of time.

Glass containers

Glass containers (see Figure 7.2) are highly desired for many products because of the customers ability to see what is in the container, that is, the product. The clarity of the glass allows for better shelf display as the customer can see what they are buying and thus give greater impulse (eye appeal) to their purchase.

Glass containers can be customized to meet the needs of the customer, that is, they may be readily blown or easily annealed to given shapes, sizes, color etc. specifications. However, glass containers are usually quite heavy, they may break easily, and light may be destructive to the retention of the quality of the food. On the other hand, glass containers are microwavable and they are reusable and, of course, 100% recyclable.

The security of the seal with glass containers maybe more difficult to accomplish than securing the seal on the metal can. Glass containers can be quite easily decorated to the customers expectations or they can be easily labeled like their metal counterpart.

Plastic materials

Plastic (a by product of the petroleum industry refining process) containers have come to the fore front for many packaging problems. They are generally inexpensive, biodegradable, relatively acceptable for heat processing, and fairly easy to fill, close and secure the seal. Some plastic containers are quite rigid which is of great value in stacking and protecting the product from crushing. Some segments of the food industry have found good use for the flexible packages and films.

Single films, that is, one layer of cellophane or mylar, may not accomplish all the desired protective properties needed, therefore, most films today are laminates of two or more films. The user should set forth the requirements for any given package, that is, the moisture barrier, sealability, gas (oxygen and odors) barrier, printability, light barrier, and/or grease resistance.

Films are being designed to modify the atmosphere in the package (MAP) and control the atmosphere in the package (CAP). Changing and controlling the atmosphere may extend the shelf life of some foods. These new films can be effective for controlling bacterial growth, prevention of staling or retention of freshness, and the control of moisture loss in many products.

The controlled atmosphere is accomplished by proper flushing with carbon dioxide and/or nitrogen after purging the oxygen from the package. The residual amount of oxygen, the added amount of nitrogen or carbon dioxide must be determined for each commodity. It should be pointed out that the atmosphere contains 78% nitrogen, 21% oxygen while carbon dioxide is only present in less than 1%. The addition of carbon dioxide prevents growth of microorganisms and it tends to stop respiration of many perishable products, thus, extending shelf life.

Controlled atmospheric (CA) storages have been in existence for some time for many fruits and vegetables and other products and this new packaging technology is only an extension of practices in vogue. The key to flexible packaging is the proper use of laminated, coextruded, and/or coated films to retain product quality during shelf life.

The traditional polymers, such as polyethyleneterephythalate (PET) films are being laminated with other films for greater control of water vapor transmission (MVTR) and oxygen transmission rates (OTR). Laminates, such as, Polyvinylidene chloride (PVDC)-coated polyester and polyethlene along with the new silica-and aluminumoxide coatings of OPP (Oriented polypropylene) films or metalized (Met) foil films may be the future for improved shelf life of many products. Many layers of films may be laminated together for packaging use, but one should remember that the cost of the package usually goes up with the number of laminated layers of materials.

TABLE 7.1 — Suggested Gases for Some Food Products

Product	% oxygen	% Nitrogen	% Carbon dioxide
Apple	2	1	97
Tomatoes	4	4	92
Cheese	—	30-0	70-100
Fish(white)	35	30	50
Fish (oily)	—	50	50
Red Meat	70	—	30
Cooked Meat	—	70	30
Poultry	—	70-0	30-100

Paper and paper products

Paper and paper products find their use in packaging in many ways. They range all the way from the Kraft bag to waxed papers, glassine to cellophane. Glassine can be coated with chocolate to make it opaque and it becomes a very good light barrier. Glassine can be grease proofed when packaging snacks or oily type products. It, also, can be coated and it can become a very good moisture barrier.

Paper comes from a renewable and biodegradable raw source, our forests. Kraft and other papers are used directly for wrapping of meats and fresh produce, they are used as secondary containers, and they are used for cartons and cases and tote boxes. However, with wood being less abundant as our forests are becoming less productive, this form of container is gradually being substituted by using plastic type materials.

Edible coatings

Edible coatings may be the answer down the road for some products, thus, the total elimination of waste, however, it is difficult for me to believe that I want to eat an edible coating wherein the product has not been packaged in some form or another. Sure we eat candies and sausage with the natural coatings of chocolate or casings, but in most cases they are protected from the elements prior to consumption. Some feel that edible coatings are nothing more than an extension of the peel, the pod, or the husk on foods.

Waxing of some commodities, such as root crops, is a good example of a man-made coating. Ice on frozen fish, collagen on meats are other present day examples. Time will tell the fate of coatings for other foods. Much effort is being put forth to make edible coatings as a protective coat or package for foods. Certainly the need is there and certainly edible coatings could be most beneficial in terms of elimination of wastes and/or their disposal problems.

One other note of great importance when considering packaging of foods and, that is, packaging of foods provide an identity to the food in terms of appropriate labeling as to nutritional values, name of the food along with its style and type, the weight of the food, the manufacturer or firm responsible for the marketing of the food, serving size, and the code. The code may be only the manufacturers code as in the case of canned products or it may be the last date of use of the product as the case for most perishable products.

Packaging of food is a necessity in this modern World and it is not expensive when one considers all the potential side effects of not packaging our food.

FIGURE 7.1 — Double Seam Terminology

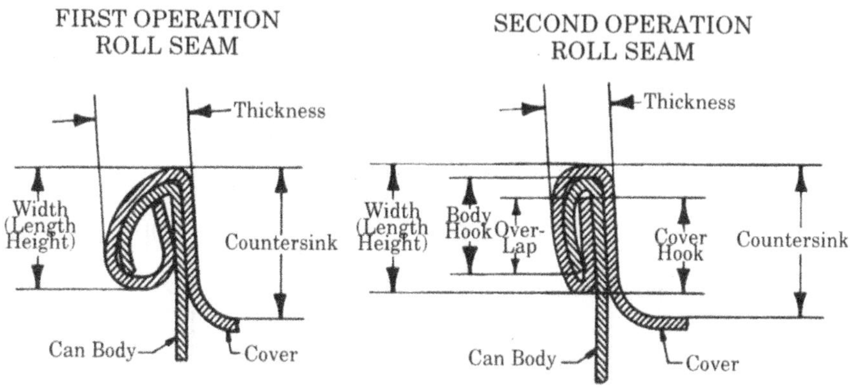

FIRST OPERATION
ROLL SEAM

SECOND OPERATION
ROLL SEAM

Minimum Measurements

Width* (not essential, if overlap
  measured optically)
Thickness*
Countersink (desirable, but not
  essential)
Body Hook*
Cover Hook* (required, if
  micrometer used)
Overlap* (essential,
  if optical system used)
Tightness* or Wrinkle

*Essential requirements.

Calculation of Overlap Length

Overlap length = CH+BH+T−W,
  where CH = cover hook
        BH = body hook
        T** = cover thickness
        W = seam width.

**In general practice, 0.010 may
be used for the tin plate thickness.

FIGURE 7.2 — Glass Container Terminology

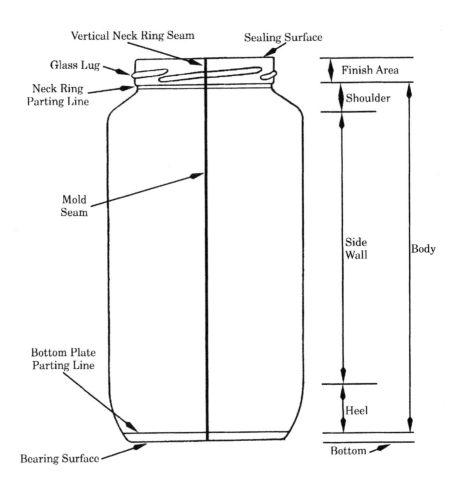

Chapter 8

FOOD PRESERVATION

"Man who say it cannot be done should not interrupt man doing it"
— Chinese Proverb

There are many methods in use today for the preservation of man's food. Some food preservation methods are very old while others are just on the horizon. The following is one way to classify food preservation methods:

1. **BAKING** — To cook food in dry heat, usually in an oven long enough to form a crust on the outer surface to protect the product from moisture pick-up. This method of food preservation is good for breads, cookies and other pastries, some snacks, but the shelf life is relatively short even with good packaging. Baking is one of the oldest methods of food preservation known to man. Bread is one of the oldest baked products and is considered the staff of life by many people.

The baking process is very complex in that enzymes are inactivated, microorganisms are destroyed, and flavor develops along with the loss of moisture. The original grains used in baking are quite tasteless, however, during the baking process the sugars and proteins are caramelized and some wonderful flavors develop. The process generally produces changes that are most desirable in texture due in part to the elastic properties of the grain proteins.

The secret of baking is to know the grain you are working with and know its textural making properties. Further, the mixing and kneading processes are fundamental in controlling texture. This step allows the enzymes and microorganisms to "work" or react and perform the desired characteristics in the given item. Further, the temperature in the baking step delegates the desired crust and color of the finished items.

In addition to bread and bread products, baked items include pies, cookies, cakes, and many snack items. Many of the baked products have fruits added to them creating problems in good crust formation and shelf life due to the high moisture in the fruits. Generally, unless frozen, baked products have a shelf life of 1 week or less for best flavor retention. However, through refrigeration and/or freezing, the shelf life may be extended considerably.

Baked products are an important part of most meals ranging from the wonderful toast for breakfast to the cake or pie treat for dinner. Baked products add color, flavor, nutrients, and appeal to any meal function. Baked products are good for you in terms of nutrition, but generally the sweets carry considerable calories.

2. **CANNING OR THERMAL PRESERVATION** — A method of preserving food whereby the food is either prepared as for table use, filled in cans, glass jars, or flexible containers, hermetically (air tight) sealed and so processed by heat to preserve the food product or it may be prepared for table use, aseptically (free of microorganisms) heated and filled aseptically into containers, hermetically sealed and cooled.

The amount of heat necessary depends on the number and kinds of microorganisms present in the food, pH of the food, consistency of the food, particulate size of the food, volume of the food (size of the container used, that is, if cooked in the container), amount of moisture initially present in the food, and the food composition.

A given process can and is calculated today for most processed items, however, due to wide variations within most canned items it is suggested that the sterility of the item be determined by inoculation and actual rates of heat penetration along with the heat resistance of the organisms that may be present. Thus, the theoretical or calculated process is tested or confirmed by the processor and/or his technologist.

In 1972 the first "Better Process Control School" was held at the University of Maryland, and today many universities utilize the Food and Drug program as developed by the National Food Processors Association. The second school was offered at The Ohio State University in 1973. In that year we offered the school twice as the enrollment far exceeded our expectations. Today that program is still being offered each year with large enrollments. The school is a good fundamental school for those that have not taken food processing in college and, even for those that are graduates, as it is a good program for building on a better understanding of the principles involved in canning technology.

Preserving food by canning is a safe method of food preservation and it is effective in retaining most of the nutrients. It, also, protects the flavor, color, and texture of the food and the resulting product is similar to what would be found when cooking the food in the home. Canned foods are nutritious, safe, convenient, and ready to eat.

The methods of preservation by canning vary widely from the still retort or pressure cooker to enormous units that agitate the food while it is being sterilized. Generally, as the cooking methods for food sterilization move from stationary retorts to mechanical systems like the Malo crateless system, to continuous rotary hydrostatic cookers, through to aseptic sterilizers and fillers the time is shortened to provide sterility and, thus, the food is cooked less in the container. Higher temperature and shorter time cooks generally produce less nutrient loss or leaching and generally produce brighter colored products. The flavor and texture may not be as good, depending on what one is comparing to. Overall, the shorter the cook time the better the texture, flavor, and color of the product. However, the time and temperature are fundamental to assure sterility.

Canned and glass packed items have been kept for several years, however, it is recommended that most canned products have a shelf life under 24 to 30 months.

3. **CONCENTRATION WITH OR WITHOUT SUGAR** — A method of food preservation by concentrating (removing the moisture from the product by boiling or vacuum concentrating) the food material with or without added sugar through heating to a solids content sufficient to protect the food from spoilage.

Concentrated products are a major source of raw materials for the secondary food processors. Concentration factories are generally located at the source of production, the raw material is prepared for consumption, generally crushed, the enzymes are inactivated, and the product is either boiled down with atmospheric cookers or under vacuum.

Atmospheric concentration goes all the way back to making maple sugar from sap where in some 40 gallons of sap are required to make 1 gallon of syrup.

Today most concentrations range from 2 to 7 fold depending on the product, ultimate usage, and concentrating equipment. Most concentrating equipment is continuous and may run for many days without stoppage due to the high volume of product in stream. Vacuum concentration is not detrimental to color or flavor loss if done at low temperatures. Wherever flavor or essences are fragile, they may be trapped and added back to the product as needed. Regardless, concentration is a highly technical aspect of food preservation and it is an economical move to preserve many food items.

Many concentrated products are on the market and they are of excellent quality. Orange concentrate that came on the market over 50 years ago is a great example of what can be done and how the consumer can and will accept a good product in concentrate form.

Concentrated products have a relatively long shelf life (2 years or more) if properly packaged and kept at relatively low temperatures.

4. **DRYING AND/OR DEHYDRATION** — One of the older methods of food preservation wherein the moisture is evaporated by exposure of the product to heat (solar, microwave or artificial) and air. It is a process of baking and drying governed by the fundamental principles of physics and physical chemistry. Drying is an energy intense process using heat applied externally or generated internally by microwave or radio frequency energy to evaporate water on the surface and drive trapped moisture to the surface. Efficient drying is only maintained by heat that penetrates to the center of the food particle forcing the moisture to the surface for evaporation or the point of lower vapor pressure area of the food particle. It takes 1000 BTU's to remove one pound of water plus additional energy to heat up the product. The heating mediums include natural or LP gas, steam, thermal oil, or electricity.

Water is essentially in 3 zones in a food particle (see Figure 8.1). The water on the surface of the particle is removed in a constant rate. Internal moisture is more difficult to remove because it is more tightly bound and is protected by the insulating effect of the already dried material closer to the surface. The water in the center is very difficult to remove and

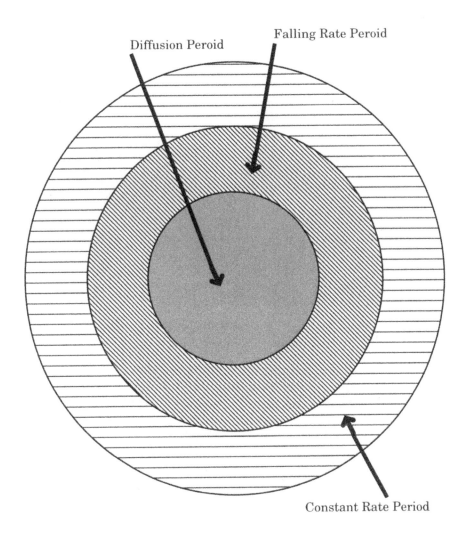

FIGURE 8.1 — Three-Zone Moisture Removal

the removal of this water is removed by diffusion. Thus, called the diffusion period of drying, that is, this water must be diffused from the center outward for evaporation to accomplish a dried product.

FIGURE 8.2 — Dry with Oscillating Feeder
(Courtesy Proctor & Swartz)

With dehydration, relative humidity is, also, controlled, thus controlling the rate of evaporation. If water is removed too fast, the food material may "case harden," that is, seal up the outer surface area so that the water cannot diffuse from the center outward. Depending on the food, the food is usually dried to less than 95% of the original moisture content.

Air is used to carry away the moisture from the surface of the product. Air is recirculated with a small volume exhausted to remove the evaporated moisture. Depending on the type of the dryer, air may be circulated through the product, over the product, or across the product. Uniformity of air flow is the most critical factor in acceptability of the drying process.

A typical drying curve is shown in Figure 8.2.

Drying of foods requires proper preparation of the raw materials as for direct consumption, that is, all vegetables must be blanched to inactivate the enzymes while most fruits maybe blanched or chemically treated to prevent the enzymes from acting to cause discoloration and off-flavors in the finished products.

Many dried products have been accepted for long periods of time, that is, spices, herbs and some sun dried products. Artificially dried foods have been well accepted since World War II. Dried foods have a long shelf life (3 to 5 years) if gas packaged in moisture vapor proof containers.

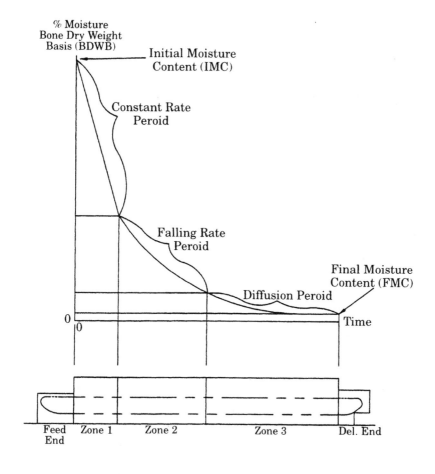

FIGURE 8.3 — Drying Curve vs. Conveyor Dryer Zoning

5. **FERMENTATION** — Fermentation is a process whereby microorganisms decompose sugars and other chemicals to acids, alcohol's, and/or carbon dioxide and water. Fermentation can be controlled by heat and/or salt to produce alcoholic or non-alcoholic products. Once the product is properly fermented, the shelf life is good for many years. As a matter of fact, many fermented products may improve with age.

Fermentation requires an understanding of the role of specific microoganisms and what substrate they can or will work on and the conditions in terms of levels of acids, sugars and salt. For example in the fermentation of many vegetables the correct level of salt or brine is a must in the control of the sequence or activity by specific microoganisms. Sometimes it is necessary to heat the fermentable product to destroy the microbes that are present prior to innoculation with specific organisms.

6. **FREEZING** — A method of food preservation developed in the 1920's with Clarence Birdseye given the credit as the "Father of Frozen Foods". The process requires the food product to be frozen quickly (usually 30 minute or less for the product to pass through the Zone of Maximum Crystallization, that is, 32–25°F or –4 to –1°C) and the frozen product must be kept at 0°F or 17.8°C. Usual shelf life is under 20 months, depending on the amount of fat in the product.

Frozen foods are a major part of the foods we eat. They generally retain the nutrients found in the original raw product and they generally have excellent color and flavor. Some people question the texture if comparing to a cooked product. That should not be a problem if the user learns to handle frozen foods in the kitchen properly.

7. **FRYING** — This is a form of drying with oil used as the heating medium. Temperatures may exceed 400°F or 204.4°C in the fryer and the products can be fried in seconds. For example potatoes having a specific gravity of 1.080 (total solids of 20.2%), with a slice thickness of 0.065" of an inch can be fried to 2% moisture content in 375°F or 190°C oil in less than 130 seconds. Temperature has a direct effect on the flavor and quality of the finished product. The product is usually fried to under 2% moisture content and the shelf life is restricted by the type

and quality of the oil used in the fryer and the packaging material. Generally for potato chips, the shelf life is less than 12 weeks while with other snacks the shelf life may exceed 20 weeks. Improvement of shelf life can be achieved by nitrogen flushing the bags after filling before closing.

8. **RADIATION** — This is a relatively new technology of food preservation having application on spices, some fruits and vegetables, potatoes in Canada, etc. The method involves the use of X-rays, Gamma rays and other sources of energy to destroy the microorganisms that cause spoilage and/or decay. Obviously, packaging should be a part of the process so that the commodity is not recontaminated.

9. **CHEMICALS: SUGAR, SALT, SPICE, ACIDS** — Many chemicals have wide application in food preservation, however, the amount of the chemical may be too excessive to make the additive practical. Acids, however, can be safely used and they can be effective in lowering the pH and making the food environment unacceptable to support the growth of public health organisms. FDA has permitted acidification of food products to aid in their preservation. Of course, sugar is part of the concentration of many products in jams, jellies, preserves, tomato sauces and ketchup. Salt as indicated above controls the rate of fermentation of cabbage and pickles.

10. **OTHER: EXTRUSION, FREEZE DRYING, SMOKING, DEHYDROFREEZING, REFRIGERATION WITH OR WITHOUT CONTROLLED ATMOSPHERE, ETC.**
    There are many other methods of food preservation and others may be forthcoming as new food preservation techniques. For example, controlled atmospheres for fruits, vegetables, meats, flowers etc. is a very practical method of extending shelf life. Using potatoes, the controlled atmosphere (CA storage) is a practical way of putting the potato to sleep for several months without detrimental effects on potato quality. As a matter of fact, CA storage extends the life of the tuber by several months and CA stored potatoes may be superior in quality for chipping and other purposes.
    One other method of food preservation that is just coming to the fore front is extrusion cooking. Here the ground product

(non-particulate) is cooked and forced by using an Archimedes screw through an orifice or opening (dies) of many shapes into the atmosphere. In so doing, the moisture is vaporized, the product is cooked and the final product expands many fold depending on the grain or vegetable or other product. The moisture is generally very low and the product has good shelf life and is quite acceptable from a flavor standpoint. Most people are not ready cooked cereal or grain eaters and, therefore, the industry flavors extruded products with cheese, spices, fruit and artificial flavors, etc. This method of food preservation is practical, safe and very efficient.

FIGURE 8.4 — Psychrometric Chart - Properties of Air And Vapor Mixture (Courtesy Proctor & Swartz)

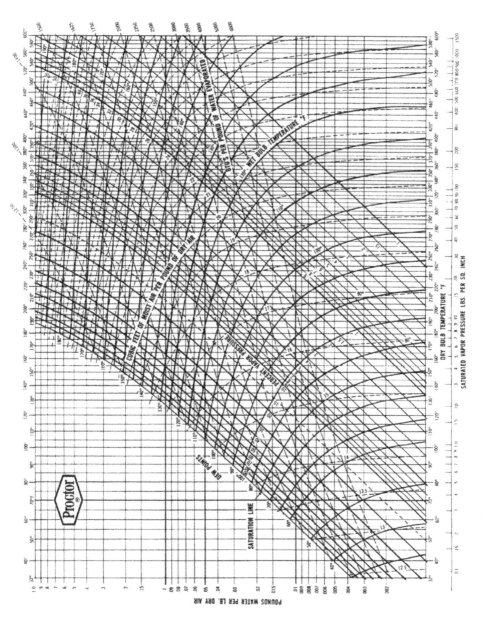

Chapter 9

MATERIALS FOR FOOD PRESERVATION

"There is no abstract art, you must always start with something"
— Pablo Picasso

Converting raw materials into value added products are the basic reason food firms are in the business. The food firm uses raw materials as their starting point in all primary processing operations. Secondary processors may start with the raw material already prepared either as a stored prepared product or they may start with a concentrated product and remanufacture forward.

Most food processing firms start with the raw material and convert it to an acceptable finished food product. In so doing, the food processor adds value to the raw material during the development of the finished product. For example, ketchup is no longer tomatoes, sauerkraut is no longer cabbage, pickles are no longer cucumbers, sausage is no longer hogs or chickens, and cereals are no longer grains. Thus, one should appreciate the food processor for his ingenuity, imagination, and creativity to develop acceptable food products ready for the user.

The food processor is in the business of adding value to the raw materials while preserving these materials for consumption. He makes them available year around, he puts them in convenient form, he preserves them, he ensures their quality, he may add nutrients to make them more acceptable, and he provides them at a reasonable price. Furthermore, he packages them and labels them for the convenience of the customer to meet their expectations all the time. Most importantly, the food processor increases jobs, creates new business, and increases the value of his products.

The foods in the market place are wide in choice and large in number. It has been reported that super stores may carry as many as 25,000 items. About one-half of these items come from the plant kingdom and other half from the animal kingdom, that is, in terms of dollars, but actually about 60 percent are of plant origin while 40 percent are of animal origin. Within any group, the raw materials vary widely. Most parts of the animal, except the bone and inards are consumed in one form or another. With plants, the seeds, fruits, leaves and stems are the primary parts of the plant used for food with the exception being the many root crops where the tubers and actual roots are consumed as such.

It is very difficult to discuss all food items as their differences are quite varied. For sake of clarity and to illustrate these differences I have chosen the potato.

The Potato

One of the most universal and widely consumed food items is the potato. It is consumed as whole baked item, or it may be sliced, mashed, diced, extruded or stripped and preserved in soups, or frozen or manufactured into various potato products including potato chips. Potato chips may be garnished or seasoned with many types of cheeses, peppers, vinegar, spices, animal and fish sauces and, yes, even ketchup.

The potato is actually an enlarged underground stem, rhizome, or stolon of the potato plant. Most tubers are found at the end of a stolon. Botanically, the tuber is a stem with eyes, that is, a true leaf scare. Figure 9.1 shows a longitudinal section of the potato. Most of the tuber is nothing more than a storage area for the starch and other carbohydrates.

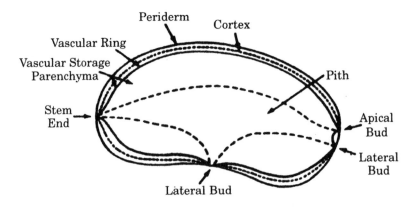

FIGURE 9.1—Longitudinal Section of the Potato
Showing All Structural Parts

The tuber is the storage organ for the nutrients manufactured in the green potato plant by combining carbon dioxide from the air and moisture from the soil to form a simple sugar. The process is called photosysthesis and requires light from the sun and the natural enzymes in the potato plant to complete the manufacture of a simple sugar with oxygen released to the atmosphere. Simple sugars (glucose and fructose) are referred to as reducing sugars and they complex and form sucrose and water. Sucrose is a non-reducing sugar and it is further complexed to form starch and give off more water. Starch is stored in the tuber and represents some 80% of the stored carbohydrates. The other carbohydrates are the sugars, acids, and flavors. The approximate composition of white potatoes is shown in Table 9.1.

TABLE 9.1—Approximate Analysis of White Potato Tubers (Taken from Talburt and Smith, 1975)

Component	Average (%)	Range (%)
Water	77.5	63.20–86.90
Total solids (dry matter)	22.5	13.10–36.80
Protein	2.0	0.70–4.60
Fat	0.1	0.02–0.96
Carbohydrate: Total	19.4	13.30–30.53
Crude Fiber	0.6	0.17–3.48
Ash	1.0	0.44–1.90

As the data in Table 9.1 indicate, the composition of any given tuber varies widely. This is due, in part, to one or more of the following:
1. Cultivar or variety as each cultivar is as different as people are.
2. Growing area and growing conditions, that is, fertilizer, water, temperature, available sunlight, and lack of stress due in great part to diseases and/or insects.
3. Maturity of the tuber at harvest and handling practices, that is, time and temperature and humidity conditions.

Any of the above may show up as tuber size differences, specific gravity or potato solids differences, shape differences, and/or types and levels of sugars in the tuber. To illustrate, see the data in

Table 9.2 on effects of tuber size on some physical constants. Note in particular the effect of tuber size on peel loss and the surface area relationship to the size of the tuber.

TABLE 9.2—Relationshipof Tuber Size to Various Physical Constants for Tubers of Different Diameters

| Physical Constants | Sphere Diameter of Potato in Inches | | | | | |
	$1\frac{1}{2}$	2	$2\frac{1}{2}$	3	$3\frac{1}{2}$	4
Surface area in square in.	7.06	12.57	20.35	28.13	39.24	50.00
Volume in cubic inches	1.76	4.19	9.16	14.13	23.82	33.51
Ratio of area to volume	4.00	3.00	2.50	2.00	1.75	1.50
App. number of tubers/lb.	19.00	8.00	5.03	2.60	1.80	1.00
Relative surface area, sq. in./lb.	133	150	111.5	73	61.5	50
$\frac{1}{16}$" peel removal in % volume loss	21	17	14.5	12	10.5	9

All of the above have a great affect on potato chip manufacture and as a result, most firms write specific contracts or sets of specifications to guide the producer to deliever what is needed and to assure the processor that he has the right raw material for his firm to operate efficiently.

A typical specification for potatoes for chipping is found in Table 9.3.

As previously stated, potato chips are made from sliced potatoes by frying the slices in oil or drying them in air. The moisture content is reduced to less than 2 percent in this process. The finished chips may be salted, seasoned or flavored prior to packaging. The package should be filled with an inert gas to protect the chip from breakage during handling and storage prior to consumption. The chip should be kept at room temperature or below and it should be packaged in a moisture vapor proof and light resistant film or container.

There are many styles of chips on the market and there are many types of seasonings and or dips in use today. The basic styles are flat chips, or wavy chips. These are due to the style of cut. The types of seasonings vary from cheese flavored to beef, poultry and fish broth flavored to vinegar, onions, and peppers. Yes, there are many exotic flavors and ketchup and mustards are finding their usage. The seasonings may be applied dry or as an emulsion spray or they may be made into dips.

TABLE 9.3—Specifications of Potatoes for the Chip Market
(by Wilbur A. Gould)

Cultivar	Norchip, Atlantic, Snowden, Gemchip and Mainechip or equivalent
Maturity	No feathering—firm skin set
Sugars: Reducing	Less than 0.15%, but prefer 0.00
Sucrose	Less than 1.5%, but prefer 0.00
Size	Minimum 2" diameter; maximum $3^1/_2$"
Shape	Round
Eye Depth	Shallow
Peel	Light in color; less than $^1/_8$" thick
Dirt	Clean and free of
External Defects	Maximum of 4% bruises, preferably 0%; no soft rot; no greening; no sprouts; no wire worm; no insect damage
Internal Defects	No hollow heart; no discoloration; no rot; no internal sprouts
Flesh Color	White to light yellow OR gold
Specific Gravity	Greater than 1.080
Total Solids	Greater than 20.2%
Chip Color	SFA 3 or lower, AGTRON (E30) (90/90) greater than 45

The yield of chips varies from 25 to 35 pounds of chips from 100 pound of raw potatoes. This wide difference in yield is due to tuber size, peel loss, slice thickness, and the original tuber dry matter content. Using potatoes with a specific gravity of 1.080 (water content of 80% and a dry matter of 20%), 5% peel loss, slice thickness of 0.060", and fried at 365 degree F. followed by 40 seconds as a finish fry at 330 degree F., one should expect a yield of 31.50 lb. of chips from 100 pounds of potatoes. By increasing the specific gravity to 1.100 (75.6% water with a dry matter of 24.4%), the yield should increase to 34.7 lb. Potato chippers use the following rule, that is, for every 0.005 specific gravity increase results in 1% increase in chip yield if all factors are kept constant. Thus, the significance of specific gravity to yield. Similar increases in yield can be obtained by decreasing the slice thickness or using potatoes with no peel loss.

The oil content of potato chips will vary with changes in the specific gravity, slice thickness, and frying parameters. The data as shown in Figure 9.2 clearly shows that higher frying temperatures and thicker slices absorb much less oil (30% v's 45%) for low frying temperatures and thin slices.

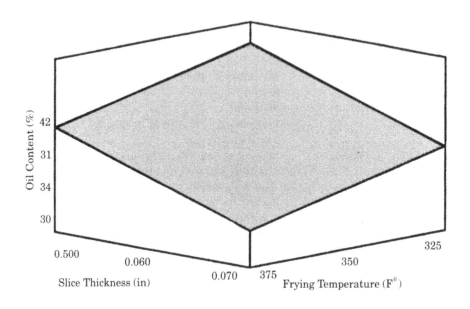

FIGURE 9.2—Effects of Slice Thickness and Fryer Temperature on the Oil Content of Regular Style Potato Chips

Some potato chip manufacturers are now making chips by drying the slices in air using commercial type dehydrators. Improved shelf life can be expected by properly blanching the slices prior to drying, but the finished chip will taste much more like a dried potato. Some chippers are spraying oil on the dried slice to attempt to simulate the flavor of the conventional chip.

The above illustrates some of the many changes for a given commodity. It all starts with specifying and using the right raw material. Anyone that states that a potato is a potato does not really know the difference among the many potato cultivars including the many inherent compositional differences affected by maturity and harvesting and handling practices.

A simple product like potato chips becomes quite complicated when attempting to provide a quality product efficiently. Raw materials do make all the differences when processing most food items.

POTATO & POTATO CHIP TERMS & TERMINOLOGY
(by Wilbur A. Gould)

BRUISE — Common defect on potatoes. Generally differentiated from other mechanical defects in that the tuber skin has not been broken.

COLD CHIPPERS — A term used to define potato cultivars which have the genetic potential to produce light-colored chips directly out of long-term storage (6 to 8 months) at 40°F (4.4°C).

CULL — A rejected product because of inferior quality.

CULTIVAR — A term used to identify a variety or new line of a given commodity. Originally used to identify numbered lines, but now includes named varieties.

DEFECT — Anything that is not wanted, an imperfection, or something that is foreign to the product.

ENZYME — A complex, mostly protein product found in all living cells that induces or speeds chemical reactions without itself being permanently altered. Enzymes must be controlled to prevent browning or oxidation of potatoes or rancidity of chips.

FABRICATED CHIPS — A style of chips wherein potato flour and/or other flours are made into a masa, extruded into a given shape, and fried or dried.

FAT — An edible mixture of triglycerides that are solid at ambient temperatures.

FIRE POINT — The temperature at which a fat or oil will burn when heated above its given temperature.

FLASH POINT — The temperature at which a fat or oil will flash when a flame is passed over the surface of the oil. This is generally some 20 to 30° below the Fire Point.

FREE FATTY ACID — The free acids formed when a fat or oil begins to break down.

GUMMING — Formation of a gum or sticky material resulting from continued heating of fats or oils. It is produced by oxidation and polymerization of the fat or oil, and represents breakdown products which collect on heated surfaces.

HYDROMETER — A floating device developed by Ora Smith and SFA to determine the specific gravity or density of potatoes.

HYDROGENATION — Hydrogenation is a process used to change a liquid oil into a semisolid or solid fat at ambient temperatures. Hydrogenation provides stability to fats, it results in higher melting points and longer shelf life without rancidity.

KETTLE CHIPS — Sliced potatoes that may or may not be washed and fried at a somewhat lower temperature than regular chips and usually are more curled due to system of frying, and may have more oil adsorbed due to longer fry time at lower temperature.

LOT — Any number of containers or loads of the same size and type which contain a product of the same type and style located in the same warehouse or conveyance, or which, under in-plant inspection or in-process, results from which consecutive production within a plant, and which is available for inspection at any one time.

LYE — A strong alkaline solution, usually Caustic Soda (sodium hydroxide) used for cleaning fryers to remove polymerized fats, etc.

MAILLARD REACTION — A dark color to chips resulting from the chemical interactions between sugars and proteins.

MAJOR DEFECT — A defect, that is, discolored area, that is up to half an inch in size.

MATURE/MATURATION — The phenomena of ripening in plants. Potatoes are considered mature when the vine is deceased and the tubers are free from feathering. Chemically, it is when the sucrose content of the enlarged tuber is at its lowest point, hopefully, less than 1.00%.

MINOR DEFECT — A defect on a potato chip that is less than a quarter-inch in size.

MOISTURE — In raw potatoes it is referred to as water content, but in chips it is referred to as moisture content. It should be below 3%, preferably less than 2%, for good shelf life.

MONOSACCHARIDE — A 6-carbon sugar, such as glucose.

OIL — An edible mixture of triglycerides that are liquid under ambient temperatures.

OXIDATION — A chemical reaction involving the addition or combination of oxygen with another chemical.

OXYGEN SCAVENGER — An antioxidant. Sometimes used in oils coming from the refiner, or it may be added to salt and seasonings or packaging materials to help prevent oxidation of the oil or the finished product.

PATHOLOGICAL DEFECTS — Defects caused by microorganisms. Examples are soft rot, scab, dry rot, blight and necrosis.

PERIDERM — The outer cells, commonly referred to as peel, of the potato. In the potato these outer cells or peel are dead cells, and provide some fiber and minerals to the consumption of same.

PEROXIDE VALUES — Fats consist of saturated and unsaturated acids in combined forms with glycerin. The unsaturated acids are susceptible to oxidation. Oxygen can add to a fatty acid to form peroxides or hydroperoxides. The peroxide value is a measure of the amount of these products.

PHYSIOLOGICAL DEFECTS — Defects caused by handling or storage practices, such as black heart, growth cracks, heat necrosis and internal sprouting.

PITH — The central area of the potato tuber or the main storage area.

POLYMERIZATION — The forming of gummy substances when frying due to oil breakdown resulting in greasy or oily finished products.

POTATO CHIPS — Thin slices of potatoes, fried in oil, and may be lightly salted. They got their start as "Saratoga Chips," named for their birthplace, the resort at Saratoga Springs, NY. In order to satisfy a difficult customer, George Crum, the Native American Chef of Moon's Lake House, prepared French fried potatoes so thinly they became mere crisps. The term crisp is still used in England for what we call chips.

RANCIDITY — An oxidative deterioration in foods containing fats whereby a typical off-odor and/or off-flavor is produced.

REDUCING SUGARS — A 6-carbon sugar like glucose or fructose which is easily oxidized. Reducing sugars react with given amino acids and cause potatoes to fry dark. Reducing sugars are formed from sucrose and should never exceed 1.5% for use in the manufacture of acceptable colored chips.

RELATIVE HUMIDITY — Ratio of water vapor present in the air to the quantity that would be present if the air were saturated at the same temperature. Potatoes should be stored with the RH greater than 90%.

SAMPLE — A representative part of something selected at random from a lot and used for inspection.

SAMPLING — The act or practice of selecting samples from a lot or load used for inspection or acceptance or rejection.

SAPONIFICATION — The hydrolysis of mono-, di- or triglycerides with caustic or alkali to form free glycerol and fatty acids in the form of soaps.

SHELF LIFE — The duration of time a product will remain acceptable to the user.

SLICE THICKNESS — The thickness of a potato slice, generally measured in thousands of an inch. The normal thickness will range from 0.050 in. to greater than 0.80 in. Thin chips, that is, less than 0.055 in., will break easier and have more oil adsorbed to the chip for a given weight of chips.

SMOKE POINT — The temperature at which a fat or oil gives off a thin continuous stream of smoke or a sign of imminent breakdown of the fat or oil.

SPECIFIC GRAVITY — A measure of total solids content of a product. Potatoes are composed of solids and water. Potatoes normally range from a total solids content of 13 to 37%, which means the water content ranges from 63 to 87%. Lower water content potatoes or higher solid content potatoes are preferred.

STARCH — A white, odorless, tasteless complex carbohydrate produced in plants as an energy source. The primary component of potatoes is starch.

SUCROSE — A 12-carbon disaccharide sugar found in potatoes and made up of one molecule of glucose and one molecule of fructose. Sucrose is chemically reduced in the potato plant to form starch.

SWEETENING — A process of starch breaking down to sucrose and in turn to glucose and fructose after a given resting period. The objective of storage is to prevent the potatoes, following harvest and storage, from sweetening.

STYLE — The characteristic or form of manufacture of potato chips, such as Regular of Flat; that is, plain slicing of the potato versus Wavy or Ridged cutting. The latter may have more oil adsorption if cut at the same thickness.

TOCOPHEROL — A naturally occuring antioxidant found in some vegetable oils that retards the onset of rancidity.

TURNOVER — The rate at which fat is used up during a frying operation. It is affected by the amount of fat adsorbed per unit of fried food, the number of units being fried during the given heating period, and the ratio of amount of fat to fried food in the fryer at one time.

TUBER — Tubers are an enlarged underground stem of the potato plant bearing buds or eyes in their axils from which sprouts and/or new plants may be formed.

Chapter 10

Management and Labor

*The game of life is not so much in holding a good hand
as playing a poor hand well.*

— H. T. Leslie

The three basic functions of Management are: (1) to establish the policy and/or goal of the firm including the objectives, mission, and value statements; (2) to make decisions for the firm; and (3) to exercise control over the firm.

POLICY

The first control over any policy is to have them. After that, it is a matter of seeing to it that they are expressed clearly; that those affected by them or responsible for their interpretation are supplied with clear cut copies; that they are kept appropriate for periodic review; they are complied with; and finally, that either rules or procedures are developed for the execution of these policies.

The objectives as part of the policy includes what the firm is trying to do, how they are attempting to execute them, who is responsible for what, where will the emphasis be placed, and when will changes, if any, be implemented. The firm should have a written mission statement clearly focusing on what it believes. Further, the firm must develop a statement of values. The value statement should proclaim the kinds of materials they will use, the process procedures, the control to the parameters that have been established, and the marketing practices that will prevail. All of the above must be in writing and it is management's responsibility to uphold these policies above everything else.

To help management make the above decisions requires very careful planning. Planning is nothing more than a precise method of accomplishing the first function of management.

DECISIONS

To make decisions, management must have good intelligent data, in other words, precise information. These data may come from peers, consultants, periodicals, reports, or observations of given operations. The more fully, systematically, and critically the information, the easier it is to arrive at the right decisions to reach the goal of the firm.

All decisions should be constantly up-dated as new information is compiled. The use of statistical techniques to make forecasting and projections is most valid. Management should become very involved with trends, new competitors, and strategies that lead to forward movements of the firm.

Decisions should never be set in concrete. They must be flexible and they must move with the times. Most importantly, all decisions must rely on analysis of all the data available.

CONTROL

In exercising control over the firm one should clearly understand the PARETO Curve, that is, any series of elements to be controlled, a selected small fraction (referred to as the vital few-20%) always accounts for a large fraction (trivial many-80%) in terms of effects. Examples of this concept are: Approximately 20% of your ingredients account for 80% of the costs, 20% of your quality characteristics account for 80% of the your customers complaints, and 20% of the product defects account for 80% of your rejections. To exercise control over any operation it is essential that constant evaluation of the process is a necessity.

MANAGER

In any successful firm, the most important part of management is a leader or manager. The manager sets the priorities. He establishes the goals. He develops the plan for the day, the tomorrow's, and the future. He provides the direction for his people, his process, his products, and his firm. He sets the tone by his leadership style. Good managers structure their jobs first and the jobs of their people on a results oriented basis. Then he structures the entire organization to obtain the desired and expected results.

Good managers concentrate on priorities and the essentials for success. Good managers delegate and hold those responsible accountable. Good managers know the performance standards of their people. Good managers rely on facts and they motivate their people to obtain the desired results. Good managers train their people on how to plan and organize to achieve the expected results. Good managers make effective decisions that can be translated into profitable actions. Good managers succeed because they run results-oriented rather than task-oriented operations. Good managers reward results not activities. Most importantly, good managers listen to reason, that is, why it should be done, why it should be done by someone else, or why it should be done in a different way.

Managers may fail if they do not maintain high morale among all subordinates, if they are unable to identify and solve problems, lack tact in working with people, inflexible, no confidence, no follow through, disloyal to the organization, associates, superiors and/or subordinates, or resists change or cannot develop a team effort or esprit de corps.

Rules for Food Plant Management

1. Promotes personal cleanliness among all employees.
2. Provide proper toilet, hand-washing facilities and lockers for all people.
3. Adoption of the Current Good Manufacturing Practices regulations and maintains a clean factory and clean equipment.
4. Rejection of all incoming contaminated materials.
5. Maintenance of proper storage temperatures and conditions.
6. Rotation of all stock and rejection of any spoiled or adulterated materials.
7. Elimination of poisonous chemicals in or near foods.
8. Assignment of cleaning and inspection of plant equipment to trained and dependable personnel.
9. Keeping buildings insect, bird, and rodent proof.
10. Keeping doors closed and windows closed or properly screened.
11. If bait boxes are used, making certain that bait boxes are locked and checked on a regular basis.

12. Removing and preventing litter accumulation in and around the buildings.
13. Storage of all food, at least, 18 inches from walls.
14. Making certain that all food being processed is suitable for human consumption.
15. Using acceptable practices to be certain that all operations are performed to protect the food from any contamination and that the food is safe for human consumption.

Keys for Leadership

Managers, supervisors, and leaders must set good examples for all to observe. Co-workers and employees want to emulate the leader, therefore, it is important that the leader is worth emulating.

People want to know the why of what they are doing, therefore any leader should explain and/or give a sense of direction for their better understanding. Most importantly, if changes are imminent, let people know what is in the offing. In other words, keep all informed.

Never give orders, give requests and make suggestions.

Always emphasize skills, not the rules. Display great enthusiasm for your work, that is, set a good example.

Praise in public and criticize in private. Always give credit when it is due. Always make new ideas welcome. Most importantly, learn to be a good listener, be respectful, courteous and be a team player.

If you are the leader, give clear directions and suggestions and always use sound judgment and pay attention to the details.

Make people feel they belong and that they are valued members of the team and firm.

The Food Plant Worker

The food plant worker is and has been the soul of the food firm. Food plant workers generally are not highly educated, but they become an essential factor in the success of the process. They are trustworthy, hard-working, intelligent (considering their general lack of much education), considerate, and reliable. Food plant workers I know are thorough and want to grow with the firm.

Some important rules for the food plant worker:
1. Always wear caps, hair nets, clean clothes and use hair restraints where appropriate.
2. No pins, buttons, jewelry, sentimental pieces, curlers, fingernail polish and/or loose attachments shall be worn at any time in the food factory.
3. No pockets above the waist line. No watches or pencils above the waist line.
4. Protective clothing and shoes should be worn at all times where appropriate.
5. Gum or tobacco chewing, smoking, and/or eating food shall be confined to designated areas.
6. Glass bottles, plastic bottles, etc. shall not be permitted in work areas.
7. Hands shall be washed and sanitized as follows:
   a. When reporting for work
   b. After a break period
   c. After smoking or eating
   d. After picking up objects from the floor
   e. After coughing, sneezing covering mouth
   f. After blowing nose
   g. After using the toilet facilities.
8. All employees must report any blemishes or break in the skin to the supervisor prior to reporting for work. Band aides, adhesives which may become loose shall not be permitted.
9. Safe personal conduct within the food plant must be strictly enforced. Running, horseplay, riding on trucks or lifts, taking shortcuts are prohibited.
10. All employees should share responsibilities to maintain lockers, dining and break areas in neat, clean, and in an orderly manner.
11. Always prevent any possible product contamination.
12. Develop and use safe practices at all times including correct use of any food additives, cleaners or other approved chemicals.

Food plant workers must become team players and be interested in knowing how to keep score. They must want to win. They must be dedicated to the task at hand. Further, they must care and always do what is right. Teams are the wave of the future and if individual employees do not want to become a team player they must seek other employment. Happy team players make for a great workplace and they can achieve so much more than unhappy employees.

The food plant worker today and in the future must be better educated and trained to handle the new equipment being developed and, in some cases, already in use. This equipment is much more sophisticated and computerized. The future worker will need extensive training in the use of the computer and understanding its application. Furthermore, much of the processes in a food factory will be computerized in terms of control, formulation, process technology and packaging. New trends will require the worker to participate in short courses, seminars and workshops.

Every food firm will spend a given percentage (0–5%) of the operating costs on training and retraining peronnel. Every firm should start today to employ college graduates who are computer literate, and computer programmable. The time has come to improve productivity and the step in the right direction starts with employing personnel that can lead a person or firm into the future.

Chapter 11

TECHNOLOGY

When there is an open mind, there will always be a frontier.
— Charles F. Kettering

The food processing industry is changing and with these changes new technologies are taking their rightful place in keeping the industry moving forward. Some of these technologies are an advancement of where we are at the present time, but they are technologies that the industry must cope with.

Quality Assurance

Quality Assurance has shifted from the laboratory to the production floor. The production worker after proper and adequate training is now fully in control of the process. This has taken years of training and sophistication of the individual unit operation. Management has removed a layer of supervision and empowered the individual worker and management holds the individual worker or his team accountable through the use of statistical data. In other words the operator of each unit operation keeps score and they always know if they and/or their team is winning.

This change in assuring quality has added much pride to the individual and his team. It is a step that motivates and allows much more freedom for the worker to do their thing and do it right at all times. Oh sure, quality assurance personnel still exist as they become the trainer and they are responsible for evaluating the data from the line and, of course, auditing of the process and the product. Quality assurance personnel spend much of their time improving the process by utilizing the data. They also work on new technologies to improve the product.

In some countries and perhaps all over the World the adoption of ISO 9000 certifications may be most important. The certification assures all of what is going on. ISO 9000 and the additional ISO's are necessary if one is into World marketing and food firms should be aware of this technology.

Food Plant Sanitation and Safety

A second technology that has helped the food preservation industry continue to grow and move forward is the modern technologies involved in keeping the plant clean and always in operation. Elbow sanitation is out and the understanding of detergents and cleaning in place (CIP) is in and most essential.

The Current Good Manufacturing Practices (CGMP) are here to stay and they should be understood by all personnel working in the food industry. For examples, in Section CFR 110.40: "The food plant equipment must be adequately cleanable and properly maintained." This implies the construction of food equipment using stainless steel. It implies free of breaks, open seams and cracks. All contact surfaces shall not impart odor, color, taste, or adulterate the food material(s). All food equipment shall be readily accessible for manual or in place cleaning. There shall not be any dead ends in any of the equipment or parts thereof. All joints and fittings shall be of sanitary design and construction. Further, all food contact surfaces shall be protected from lubricants.

Safety has been greatly enhanced through modern sanitary practices. Food safety is a full time operation and requires a well defined program with continuous surveillance. It takes team work to keep a plant clean and good communication among all employees. Everyone is important when it comes to sanitation and a well kept clean plant.

Hazard Analysis Critical Control Point (HACCP)

Modern sanitation programs include the identification of the potential hazards and the establishment of critical control points followed by verification procedures that assure everything is as it is supposed to be. HACCP (Hazard Analysis Critical Control Point) is the name of this effort and it is a blessing in disguise for many firms that lacked proper control of potential hazards.

Hazard analysis has been defined as the identification of sensitive ingredients, critical process points, and relevant human factors as they affect product safety. Critical control points are those process determiners where loss of control would result in an unacceptable product safety risk. HACCP is a systematic approach to be used in food production as a means of food safety from harvest to consumption.

There are 5 preliminary steps to developing a HACCP program and 7 steps that correspond with the 7 principles of HACCP. The 5 preliminary steps include: (1) Endorsement and support by management; (2) Selection of a HACCP coordinator to manage the program; (3) Development of a HACCP team (Plant Engineer, Production Manager, Sanitation Manager, Quality Assurance Manager, and Food Microbiologist or Technologist); (4) Identification of each product (formulation and ingredients, how it is processed and distributed, and potential abuse, if any, by the consumer(s); and (5) a Process Flow diagram.

The 7 Principles include: Principle 1, the conduction of a Hazard Analysis. This is basically an observation of the existing process line, employee practices and how the product is handled. Of course, this is done for every product.

Principle 2 is the identification of the Critical Control Points (CCP), that is, the steps in the process at which preventive measures can be applied to prevent, eliminate or reduce the potential hazard(s).

Principle 3 is the actual establishment of the Critical Limits (CL) for each Critical Control Point(s). These CL's are expressed by numbers, that is, time, temperature, pH, etc.

Principle 4 is the establishment of Monitoring Procedures, that is, a planned sequence of observations or measurements to assess whether a CCP is under control and to produce an accurate record for future use in the verification step.

Principle 5 is the establishment of corrective actions for those times when monitoring indicates a failure to comply with the critical limits. Corrective actions require the determination of non-compliance, the cause(s) of non-compliance, correction of the cause to be certain that it is back under control, and the records to prove that the corrective action was adopted.

Principle 6 is the establishment of record keeping principles that document the HACCP system. My proposed record keeping systems are shown in the following four forms:

Form 11.1 is the identification of any potential hazards, their preventive measures, and a description of the CCP.

Form 11.2 is the Monitoring/Corrective Action form to show what, when, where, how and who of the monitoring procedures for each CCP

Form 11.3 is the verification record keeping form to show that the verification procedure was completed for each CCP by the process step.

The above three forms are developed and maintained by the operators.

Form 11.4 is a master form and developed by the HACCP Coordinator to summarize exactly what is going on relative each product.

(All four forms can be found at the end of chapter.)

Flow Chart (Figure 11.1) for Potato Chips shows the specific steps in the manufacture of potato chips and all the potential hazards.

These completed four forms and the Flow Chart provide a simple and methodical application of appropriate science and technology to plan, control, and document the safe production of a given product. They provide the data to assure a safe process from given raw materials in an appropriate environment to eliminate the testing of the final product from a safety standpoint. Thus, HACCP is prevention of hazards in the product, but, it all depends on people. People must execute what is intended and expected. HACCP is pro-active not reactive. It really gives emphasis to what we should have been doing all along.

HACCP should be welcomed by all food firms and it should be a part of their effort to assure the consumer of safe and wholesome foods. The records may take time, but they also may be most valuable if the product is not up to specification. Physical, chemical or microbiological hazards can be controlled and it behooves this industry to adopt HACCP and the many ramifications thereof to protect them from potential lawsuits and to assure the customer that your product is safe and wholesome.

The responsibilities for sanitation and food safety is a direct responsibility of management and the authority to carry safety and cleanliness out must be assigned to responsible person(s). It is the most important job in a food firm and management must make it that way.

FIGURE 11.1 — Unit Operations in Chip Manufacture

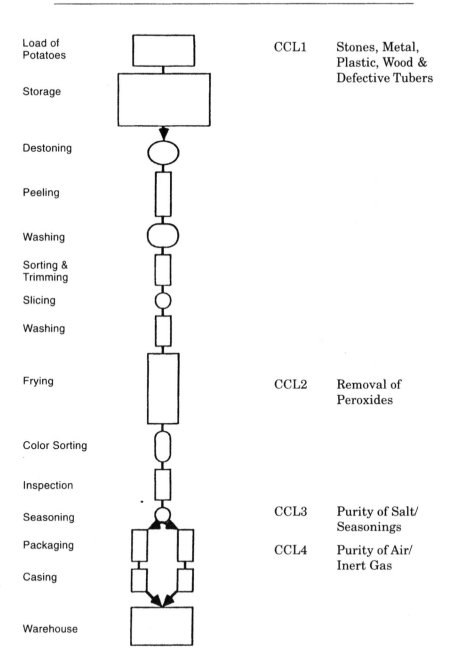

Load of Potatoes	CCL1	Stones, Metal, Plastic, Wood & Defective Tubers
Storage		
Destoning		
Peeling		
Washing		
Sorting & Trimming		
Slicing		
Washing		
Frying	CCL2	Removal of Peroxides
Color Sorting		
Inspection		
Seasoning	CCL3	Purity of Salt/ Seasonings
Packaging	CCL4	Purity of Air/ Inert Gas
Casing		
Warehouse		

Nutrition and Nutritional Labeling

Another technological program is nutrition, nutritional labeling, and the understanding of the many nutrients that may be in our foods. Nutritional labeling of foods may be costly and time consuming, but consumers want to know what they are eating and this information may be most valuable in terms of customers who want to continue to buy your products. The public has a right to know what you are putting into your packages. The label is your silent sales person and it behooves every purveyor of food to do it right all the time. Sharing this information may not always be understood by the buyer, but they can learn too and take advantage of your efforts.

New Products and New Product Development

New products are being engineered every day using precise formulations for uniform and high quality new foods. Many, many new products come on the market without adequate testing and it behooves those responsible for this technology to develop products that meet the customers demands. Utilizing market tests, focus groups, and panels must be improved to save the many dollars wasted when developing products that do not meet the demand of any potential customers.

The development of new products to meet a salesmans ideas or something I always wanted to do will not work in the future. It must be a concerted effort of many people including the food scientist, the food engineer, the food technologist, the process development technologist, the marketing personnel and many others. It is no longer a one man show. New products are needed, but copying old products to stay in line is not doing much for the future of this industry. New products must start by understanding the scientific method and its many ramifications, that is,

1. What is the type of product needed? This requires identification of the product, the precise objectives of why we want to develop this new item. We must set forth our hypothesis or our assumption that we need a new product that we have identified.
2. Find out what has been done to-date. That is, is there any published information about this new item, are there any patents, are products already in the market place?

3. How will we proceed or what will be our approach, in other words, how do we do it.
4. If the above three steps tell us to go forward, we must collect data or gather some facts, in other words work our plan and develop what we think is needed.
5. This step is most important and too often not fully utilized. We must learn to interpret our data to see if we did what we said we set out to do. That is, did we fullfill our objectives or did we test our hypothesis.
6. We must draw conclusions based on our findings, we must develop recommendations based on the facts, and finally we must summarize our efforts for this potential new item and how it might fit into our line.

New products are what keeps many firms moving forward, but they need to be clearly identified and researched to assure that this is what we want to and can do.

Records

For every technology, we must learn to develop a clear set of records (observations and measurements). Records must be accurate and reflect the actual situation or operating conditions at the time specified. All records must be in ink (ball point is acceptable), they must be signed by the operator and initialed by the supervisor or another operator. All records must be on proper forms and they should be kept for the life of the product unless otherwise indicated by the government.

Some examples of types of records include the following:
- Quality and condition of all incoming materials
- Suppliers Guarantees or Certificates
- Actual operating conditions for each Unit Operation including Statistical Process Control Charts
- Fill Weights, Vacuums, Codes and Closure Conditions
- Initial Temperatures (IT) for canned products
- Process or Cook records, that is, sterilization times and temperatures and conditions of departures
- Where applicable, pH, and other critical factors intended to ensure a safe product
- Quality Control, Evaluation and Audit reports
- HACCP records including verification record

Records are needed to keep management fully informed. These include the above plus warehouse and shipping forms. Records provide information to assure compliance with governmental regulations. Records provide evidence of safe operations in the production and processing of safe foods. Records indicate potential hazards in the food operation and the identification of all critical control points and the control of the process to given specifications during production and processing. Records verify the adequacy of the process, the integrity of the containers, and the quality assurance of the produced foods. Records identify through proper coding of all lots manufactured by the firm. Records provide the needed data to deduct the efficiency during any given production period. No firm should ever manufacture a food product without detailed records.

Ideally, records today may be developed and accumulated by using computers at given work stations and throughout the operation. Thus, the modern manager has the records as the product is being processed. He can observe the data, make recommendations to adjust, and he can observe what is going on at all times.

Some detailed help

The above technologies are some of the many facets of concern to the modern day food processor. I have published details of many of these in the following books all from CTI Publications. I would hope every firm would have these books to go with this book and, of course, my complete library to help your firm grow.

I have not reproduced the Federal Food Drug and Cosmetic Act of 1938 and its many Amendments, the CGMP Regulation of 1969 with revision in 1986, CFR 113 and 114 Low Acid Foods and Acidified Foods published in 1979, nor the new HACCP Regulation (Initiated in 1970's with adoption to seafood in 1997. These are all available from the U.S. Government Printing Office or in a convenient form in one volume from *The Almanac*.

Gould, Wilbur A. 1991. *Research and Development Guidelines for the Food Industries*. 176 pages.

Gould, Wilbur A. 1992. *Total Quality Management for The Food Industries*. 164 pages.

Gould, Wilbur A. and Ronald W. Gould. 1993. *Total Quality Assurance*. 464 pages, 2nd Ed.

Gould, Wilbur A. 1994. *Curent Good Manufacturing Practices Food Plant Sanitation*. 400 pages, 2nd Ed.

Gould, Wilbur A. 1994. *Human Resource Development for the Food Industries*. 194 pages.

Gould, Wilbur A. 1995. *Glossary for the Food Industries*. 195 pages, 2nd Ed.

Gould, Wilbur A. 1996. *Unit Operations for the Food Industries*. 186 pages.

FORM 11.1 — HACCP PROCESS STEPS/HAZARDS/PREVENTIVE MEASURES/CCP's

Process Steps	Potential Hazards	Preventive Measures	Critical Control Point/ Description

FORM 11.2 — HACCP MONITORING/CORRECTIVE ACTIONS

Process Step/CCP	Critical Limits	Monitoring Procedures				Corrective Actions
		What	How	When	Where	Who	

FORM 11.3 — HACCP RECORD-KEEPING/VERIFICATION

Process Step/CCP	Records	Verification Procedures

FORM 11.4 — HACCP MASTER RECORD SHEET BY PRODUCT

Process	CCP	Hazard	Critical Limit	Monitor	When	Who	Corrective Action	Documented	Verification

Chapter 12

SOME FUTURE CONSIDERATIONS, CONSTRAINTS & CONCERNS

The best preparation for tomorrow is to do today's work superbly well.

— Sir William Oliver

No one knows the future of food production and food processing or what lies ahead. However, there are some considerations and constraints of the processing and technology fundamentals of our food.

First, people must eat and we must do a much better job of preserving the food that we produce and process just for survival. The world population's continued growth is a reality. What was good enough yesterday is not going to be good enough for today and/or in the future. We need to continue to lead the World in food preservation methods and practices. We need to provide more processed foods at a reasonable price for the under privileged and the under developed parts of the World.

We need to understand the benefits of genetic engineering, the technologies of food processing and preservation, and we need to understand the use of our knowledge of new food developments. We, also, need to better utilize the food that we have. We waste too much of what is produced. We need to develop a better understanding of our many ethnic cultures. Further, we need to develop better systems of protecting our food by utilizing newer and better packaging methods.

Most importantly, we need to control microorganisms and totally wipe out the many diseases of our plants and animals. We must have a much better control of the many types of insects, rodents and birds that destroy our foods.

Opportunities do exist for the new food technologist, scientist, engineer, and chemist to step forward and make their contributions. Yes, even the business graduate can contribute by helping control costs, productivity, and the development of better markets for our efforts.

Every food processor should be building for the future and have definite answers to the following questions:

1. *Have we developed a POLICY statement* and communicated this along with our MISSION, VISION, and VALUES to all our employees, our peers, our customers, and the community at large. These statements must elucidate our practices for quality, food processing systems, technology in terms of types and styles and amounts of various ingredients, our packaging efforts including labeling, and most importantly our coding system as to pull date or last date to use. Included in all of these, we must state our strategic plan in terms of leadership for the tomorrow's, in terms of where we want to be as a firm, and in terms of what we want to be producing in the future. These statements help our employees understand our principles of operation so that they can become dedicated, more loyal, and enthused employees. Communication from the top down eliminates the grapevine and insures trust, confidence, and belief in the firm and its management. Some call this good old fashion planning and this is really what it is. In other words, I have a dream and I want to share it with my employees, my peers, and my customers. Most importantly, I want to share it in my community because I believe in the tomorrow.

2. *Have we established a BASELINE?* How do we compare with our competitors, in other words, AND, yes, do we have a BENCHMARK? This is all most fundamental and the old fashion "cutting bee" is so appropriate today to know where you stand. Maybe you are trying to be too perfect or maybe you are so inferior that no one really wants your best efforts.

3. *Do we have knowledge of our EFFICIENCY and PRODUC-TIVITY?* Are we making appropriate MEASUREMENTS to know what we are doing? What is our RECOVERY and are our YIELDS appropriate for this industry or are we wasting too much material in production or processing? Do we have records of our DOWN TIME and the causes of same. What

do we know about our people? Are some always off sick, with physical or mental health problems? What about our SAFETY record, are we paying excessive premiums?

4. *Do we understand the causes of VARIATION*, that is, the variables of Materials, Manpower, Methods, Machinery, and Environment? Do we know whether these sources of variation are inherent to the system that we are using , or to our suppliers and their materials, or the different people working on each unit operation, or the unit operation or machine itself? Are we keeping score by having historical data to understand when, where, why, what, who, and how of our variations or the many problems that we tend to ignore? Do we have any idea or realization of the opportunities that exist by reducing each of these variables as the data dictate? These are all management decisions, but any manager must have facts to work with for solving the problems and eliminating the variables of production for success.

5. *Does our firm understand and put into practice some of the many PARADIGM shifts* that are taking place in the food industry today, such as,
   - Tall and rigid structures Vs being flat and flexible
   - Controlled management Vs committed leadership
   - Wealth exploited Vs wealth creating
   - Command decisions Vs consensus decisions
   - Task focused Vs consumer focused
   - Experts and labor Vs all experts
   - Individual worker Vs team worker
   - Record keeping Vs score keeping
   - One right way Vs continuous improvement

6. *Are our HUMAN RESOURCES bringing the best out of our people* or do our people need more training to cope with the newer technologies, changing times, or improved methods of processing? Are we providing our people with seminars, workshops, videos, computers, and networking including plant visitations to improve their knowledge of our needs?

7. *Do we understand the newer REGULATIONS and are we keeping up to date?* Have we assigned one of our people to follow the changes that are announced almost daily? Are we sharing this information with those that are concerned?

Some Constraints

There are fewer firms in the industry today than 50 years ago. The present firms in many cases are larger, more efficient, superior in productivity, and give their utmost to produce high quality, safe food, and retention of the most in nutrients. Many of the present firms are the result of mergers, buyouts or an outgrowth of previous wars and conflicts. The food business requires committed leaders, it requires capital, and it requires entrepreneurial spirit. It takes many dollars to operate a food firm and it takes technical know how along with business accumen. Yet, one must remember all big firms today were little ones yesterday. The day of the family operation still exists and the ones I know are profitable, operating with a sense of direction toward the future, and they still have that pioneering spirit and leadership.

Risk is a major constraint, but I have always believed that if you do a good thing well, good will come to you. You don't have to run and hide, but you must understand the risk of the weather, the risk of the short season, the lack of control of raw materials due to Mother Nature, the lack of assurance of labor, and the non-interest in buyers. If you have a good idea, a good product, and you have tested it thoroughly, why not make your move? All of these risks occur in whatever you do. So, muster up your entrepreneurial spirit, make your plans, and start to work your plan today.

There are many new changes that should excite the new entrepreneur such as:

Better methods of producing, harvesting and handling crops today. This may all start with the utilization of the existing knowledge of genetic engineering and our better understanding of genes and their place within the plant or animal.

A great technology that has major impact on starting a new food processing operation is learning to utilize bulk stored products. These products are a natural for anyone wanting to invest in the food business as much less capital is required to enter and to operate the business.

Newer technologies of food preservation, such as acidification, aseptic packaging, extrusion of many products into long term shelf life items, and the use of Ohmic and other new sterilizing techniques to improve flavor, color, shelf life.

Without question, the influence of the CHEF in the development of new flavors, new textures, and foods with greater appeal are on the horizon. Some of these influences may improve a line of products or create a whole new line of products. Remember, value added is the name of the future for this industry.

One major concern for any employee or potential employee in the food business is an understanding that for every food fallacy there is a food fact. You must understand this as why the food faddist attempts to undermine public confidence about myths of DEPLETE SOILS, HEALTH FOODS, FOOD ADDITIVES, PESTICIDES, and yes, PROCESSED FOODS. We are in an industry that has kept Americans the healthiest people in the world. We are the best fed, and we have plenty of food for all our people.

The food faddist attacks government agencies and their people, they scare others of us about their quick cures, their pills, their health plans, and even their literature. Food faddists always have something to sell, usually at a tremendous profit using so called "case" histories to promote the cure. The food supplements they offer may be very dangerous if used excessively. Food faddists have misused authoritative facts and information and endangered our public health.

Each of us in the food industry has a major responsibility to obtain the facts about our food and food production and food processing, what we do, how we do it, and why we do it. We should all be very proud to extol these facts at every opportunity. We should be the spokesperson for this great industry and we should speak at every opportunity that comes along to tell our story the way it should be told. Who better if not us?

The food processing industry is a great industry. We do much for the consumer. We allow two or more in a family to work, as the housewife of today has very little time to spend in the kitchen. She should be free to work, play, and do the things she wants to do rather than work over a hot stove all day.

As a matter of fact most housewives, cooks and families really love the processed foods industry for the saving of their time, for the many wonderful meals we put together, for the convenience, for the nutrition we deliver, and for the goodness in flavor and appearance of the food we provide.

We are in the business to please and repeat business is our motto as we have a good reputation to uphold all the time.

INDEX

NOTES

Additional Titles From CTI Publications

FOOD PRODUCTION/MANAGEMENT - Editorially serves those in the Canning, Glasspacking, Freezing and Aseptic Packaged Food Industries. Editorial topics cover the range of Basic Management Policies, from the growing of the Raw Products through Processing, Production and Distribution. (Monthly Magazine). ISSN: 0191-6181.

CURRENT GOOD MANUFACTURING PRACTICES, FOOD PLANT SANITATION - This work covers all CGMP's as prescribed by the United States Department of Agriculture, Food and Drug Administration, as it applies to food processing and manufacturing. The reader is guided through the CGMP's and provided with various plans and sanitation controls. ISBN: 0-93002721-3.

GLOSSARY FOR THE FOOD INDUSTRIES - 2nd Edition, is a definitive list of food abbreviations, terms, terminologies and acronyms. Also included are 26 handy reference tables and charts for the food industry. ISBN:0-930027-23-X.

A COMPLETE COURSE IN CANNING, 13TH EDITION - A technical reference guide and text book for food plant managers, product research & development specialists, students of food technology, food brokers, technical salespeople, food equipment manufacturers and food industry suppliers. The most comprehensive volumes published on canned foods - Thousands of topics covered. ISBN: 0-930027-25-6.

RESEARCH & DEVELOPMENT GUIDELINES FOR THE FOOD INDUSTRIES - Is a compilation of all Research and Development principles and objectives. Easily understood by the student or the professional, this text is a practical "How To Do It and Why To Do It" reference. ISBN: 0-930027-17-5.

TOMATO PRODUCTION, PROCESSING & TECHNOLOGY - 3rd Edition, is a book needed by all tomato and tomato products packers, growers, or anyone involved or interested in packing, processing, and production of tomatoes and tomato products. ISBN: 0-930027-18-3.

TOTAL QUALITY ASSURANCE FOR THE FOOD INDUSTRIES - 2nd Edition - The only answer to guide a food firm, its people, its quality of products, and improve its productivity and provide that service, that food product, and that expectation that the customer wants. Every firm that endorses, resources, and practices a Total Quality Management program will find great and meaningful accomplishments today and in the immediate future. TQA will help you to more than meet your competition and build your bottom line. ISBN: 0-930027-20-5.

TOTAL QUALITY MANAGEMENT FOR THE FOOD INDUSTRIES - Is a complete interactive instruction book, easily followed, yet technically complete for the advanced Food Manager. TQM is the answer to guide a food firm, its people, its quality of products, and improve its productivity. It's the right step to achieve excellence and the development of satisfied customers, as well as build your bottom line. ISBN: 0-930027-19-1.

UNIT OPERATIONS FOR THE FOOD INDUSTRIES - This food processing operations book is a must reference for all industry individuals who need to draw on the newer technologies that are emerging in the food industry. Over 100 figures and tables. ISBN: 0-930027-29-9.

For a brochure or further information please contact:

CTI Publications, Inc.

Please See Front Of Book For Complete Address, Phone and FAX Numbers

Your Global Source for Technical Reference Books For the Food Processing Industry

Printed and bound by CPI Group (UK) Ltd, Croydon, CR0 4YY

03/10/2024

01040436-0020